Oxford
International
Primary
Science

Terry Hudson

Alan Haigh

Debbie Roberts

Geraldine Shaw

Language consultants:
John McMahon
Liz McMahon

4

OXFORD
UNIVERSITY PRESS

OXFORD
UNIVERSITY PRESS

Great Clarendon Street, Oxford, OX2 6DP, United Kingdom

Oxford University Press is a department of the University of Oxford. It furthers the University's objective of excellence in research, scholarship, and education by publishing worldwide. Oxford is a registered trade mark of Oxford University Press in the UK and in certain other countries

British Library Cataloguing in Publication Data
Data available

978-0-19-839480-8

10 9

Paper used in the production of this book is a natural, recyclable product made from wood grown in sustainable forests. The manufacturing process conforms to the environmental regulations of the country of origin.

Printed in India by Manipal Technologies Limited

The questions, example answers, marks awarded and comments that appear in this book were written by the author(s).
In examination, the way marks would be awarded to answers like these may be different.

Acknowledgements

The publishers would like to thank the following for permissions to use their photographs:

Cover photo: Science Picture Co./Corbis, P4_5: BlueOrange Studio/Fotolia, P8a: Omika/Fotolia, P8b: Fotolia, P8c: Fotolia, P8d: Sergio Pitamitz/Corbis/Image Library, P8e: Fotolia, P8f: Shutterstock, P10a: Jgi/Oup, P10b: Trevor Driscoll, P11a: Peter Anderson/Dorling Kindersley/Getty Images, P11b: Marlon Lopez MMG1 Design/Shutterstock, P14a: Emmy-Images/iStock. Com, P14b Grzegorz Placzek/Shutterstock, P14c: Stacy Barnett/Shutterstock, P15a: Nick Veasey/Untitled X-Ray/Getty Images, P15b: Srdjan Draskovic/Shutterstock, P15c: Vilainecrevette/ Shutterstock, P15d: Amy Johansson/Shutterstock, P15e: Shutterstock, P16: Photodisc/OUP, P20: Photodisc/OUP, P22a: Mark Sykes/Science Photo Library, P22b: Stephanie Rabemiafara/ Art in All of Us/Corbis/Image Library, P22c: Miriam Doerr/Shutterstock, P23: Zeljko Radojko/Shutterstock, P25: Zeljko Radojko/Shutterstock, P26: WaterFrame/Alamy, P27a: Charles D. Winters/Science Photo Library, P27b: Discpicture/Shutterstock, P28: Ethan Daniels/Shutterstock, P30: Nidal Yunus, P31a: Madlen/Shutterstock, P31b: Roxana Bashyrova/Shutterstock, P31c: Koosen/Shutterstock, P32a: Africa Studio/Shutterstock, P32b: Galyna Andrushko/Shutterstock, P32c: Edyta Linek/Dreamstime.Com, P32d: Minerva Studio/Shutterstock, P33: Rickwpeters/Dreamstime.Com, P34a: Tatiana Morozova/Dreamstime.Com, P34b: iStock.com, P35: Bhaskar Chandra, P36a: Steshkin Yevgeniy/Shutterstock, P36b: iStock.com, P37a: Shutterstock, P37b: Gyro Photography/Amanaimages/Corbis/Image Library, P38a: RyersonClark/iStock.com, P38b: Ria Novosti/Science Photo Library, P40a: NASA, P40b: Richard Wear/ Design Pics/Corbis/Imaage Library, P41a: Alina G/Shutterstock, P41b: Sheff/Shutterstock, P42a: Arangan Ananth/Shutterstock, P42b: Sujin Jetkasettakorn/123RF, P42c: Fotolia, P42d: Fotolia, P43: Elfi Kluck/Photolibrary/Getty Images, P44: Guy Sagi/Dreamstime.com, P46: Www.Railphotolibrary.Com, P47: Geoff Renner/Robert Harding World Imagery/Corbis/Image Library ,P48a: Singhsomendra/Dreamstime.com, p48b: Jason Orender, P48c: Shutterstockm, P50a: Ethan Daniels/Shutterstock, P50b: Sheff/Shutterstock, P50c: Discpicture/Shutterstock, P51a: Sujin Jetkasettakorn/123RF, P51b: Elfi Kluck/Photolibrary/Getty Images, P52_53: Nasa/Ouop, P53a: Babak Tafreshi, Twan/Science Photo Library, P53b: Gavin Hellier/Robert Harding World Imagery/Corbis/Image Library, P53c: Doug Martin/Photo Researchers/Getty Images, P54: BanksPhotos/iStock.com, P55: Tischenko Irina/Shutterstock, P58a: Veniamin Kraskov/ Shutterstock, P58b: Ingram Publishing/Superstock Limited/Oup, P58c: Shutterstock, P58d: Terekhov Igor/Shutterstock, P58e: Shutterstock, P58f: Shutterstock, P58g: Shutterstock, P58h: Dennis Kitchen Studio, Inc/Oup, P59: Istock.Com, P60a: Image Source/Oup, P60b: Mmaxer/Shutterstock, P60c: Paul Rapson/Science Photo Library, P61a: Imaginechina/Corbis/Image Library, P61b: Melba Photo Agency/Oup, P61c: Kletr/Shutterstock, P61d: Fotocrisis/Shutterstock, P61e: CreativeAct - Technology Series/Oup, P63a: Tischenko Irina/Shutterstock, P63b: Imaginechina/Corbis/Image Library, P64a: Iakov Filimonov/Shutterstock/Oup, P64b: Pu Su Lan/Shutterstock, P64c: Corbis, P65a: Corbis / Digital Stock/Oup, P65b: Image Source/Oup, P66: Emma Bradshaw, P67a: Martyn F. Chillmaid/Science Photo Library, P67b: Emma Bradshaw, P67c: Oleg Znamenskiy/Shutterstock, P67d: Roberts Ratuts/Dreamstime.Com , P67e: Bernard Bisson/Sygma/Corbis/Image Library, P68a: Shutterstock, P68b: Fabio Pupin/Visuals Unlimited, Inc. /Science Photo Library, P68c: Amazon-Images/Oup, P68d: Leo Shoot/Shutterstock, P70a: Shutterstock, P70b: Anton Petrus/Shutterstock, P72: Rob Reijnen/ Foto Natura/Minden Pictures/Corbis/Image Library, P73: Gentoo Multimedia Limited/Shutterstock, P74a: Katrina Brown/Oup, P74b: Oleksandr Berezko/Shutterstock, P75: Thinkstock/Oup, P76a: Shem Compion/Gallo Images/Getty Images, P76b: Anke Van Wyk/Shutterstock, P76c: Mark O'Shea/Nhpa/ Photoshot, P77a: Fedor Selivanov/Shutterstock, P77b: Shutterstock, P77c: Peter Jilek/Shutterstock P77d: Beth Swanson/Shutterstock, P78: Shutterstock, P79a: Holly Kuchera/Shutterstock, P79b: Sari Oneal/Shutterstock, P79c: Shutterstock, P79d: Shutterstock, P79e: Martin Fowler/Shutterstock, P79f: Sue Robinson/Shutterstock, P79g: Eduard Kyslynskyy/Shutterstock, 79h: Vittorio Bruno/Shutterstock, P79i: Natali Glado/Shutterstock, P79j: Shutterstock, P79k: Shutterstock, P79l: Robert Palmer/Shutterstock, P79m: Shutterstock, P80a: Shutterstock, P80b Shutterstock, P80c: Shutterstock, P80d: Stephen Alwarez/Shutterstock, P80e: Jaroslaw Saternus/Shutterstock, P80f: Shutterstock, P80g: Shutterstock, P81a: Shutterstock, P81b: Anne Kitzman/Shutterstock, P81c: Shutterstock, P81d: Itsabreeze Photography/Flickr/Getty Images, P81e: Jeff Foott/Science Faction/Corbis, P82a: Lee Celano/Corbis, P82b: Ricardo Figueredo/ Corbis, P84: Shutterstock, P85: Punit Paranjpe/Corbis, P86: Karen Kasmauski/Science Faction/Corbis/Image Library, P87: Maximilian Stock Ltd./Science Faction/Corbis/Image Library, P88: Joe Carini/Design Pics/Corbis/Image Library, P89: Dr Juerg Alean/Science Photo Library, P90a: Shutterstock, P90b: Joel W. Rogers/Corbis/Image Library, P91a: Stanislav Komogorov/ Shutterstock, P91b: Vibrant Image Studio/Shutterstock, P91c: Shutterstock, P91d: Sky Light Pictures/Shutterstock, P91e: Shutterstock, P93a: Michael S. Yamashita/Corbis/Image Library, P93b: Ilona Ignatova/Shutterstock, 93c: Christophe Testi/Shutterstock, P94a: Emma Bradshaw, P94b: Emma Bradshaw, P94c: Martyn F. Chillmaid/Science Photo Library, P94d: Oleg Znamenskiy/Shutterstock, P96_97: Rob Howard/Corbis/Image Library, P96: Doug Martin/Science Photo Library, P97a: Ely Solano/Shutterstock, P97b: Masterfile, P98: Martyn F. Cillmaid/ Science Photo Library, P99a: Shutterstock, P99b: Tatiana Popova/Shutterstock, P101: Mark Scheuern / Alamy, P102: Shutterstock, P105: Vadim Shelgachev/Shutterstock, P106a: Doug Martin/Science Photo Library, P106b: Andrew Lambert Photohraphy/Science Photo Library, P106c: Lasse Kristensen/Shutterstock, P112_113: Shutterstock, P112: Ingram/Oup, P113a: Corbis, P113b: Eye Ubiquitous/Rex, P113c: Milos Luzanin/Shutterstock, P114: Jon Hicks/Corbis/Image Library, P115a: Andrew Lambert Photohraphy/Science Photo Library, P115b: Music Alan King/Oup, P116a: Andrew Barker/Shutterstock, P116b: Dave Allen Photography/Shutterstock, P116c: Shutterstock, P118: Ole Graf/Corbis/Image Library, P119a: Caleon, I. S., Subramaniam, R. & Regaya, M. H. P. (2013). Revisiting Bell-Jar Demonstration. Physics Education, 48 (2), 247-251, P119b: Visual Photos / Science Photo Library Rf / Nasa, P120: Samuel Micut/Shutterstock, P123a: Ian Boddy/Science Photo Library, P123b: Shutterstock, P124: Monty Rakusen/Cultura/Corbis/Image Library, P125: Ben Bryant, P126a: Oliver Benn/The Image Bank/Getty Images, P126b: Shutterstock, P127: Jeremy Horner/Corbis/Image Library, P128: Wikihow, P129: Shutterstock, P130a: Moodboard/Oup, P130b: Tim Graham/Getty Images, P131: Lourens Smak / Alamy, P132: Andrew Lambert Photohraphy/Science Photo Library, P133: Monty Rakusen/Culture/Getty Images

Although we have made every effort to trace and contact all copyright holders before publication this has not been possible in all cases. If notified, the publisher will rectify any errors or omissions at the earliest opportunity.

Links to third party websites are provided by Oxford in good faith and for information only. Oxford disclaims any responsibility for the materials contained in any third party website referenced in this work.

The questions, example answers, marks awarded and comments that appear in this book were written by the author(s). In examination, the way marks would be awarded to answers like these may be different.

Contents

How to be a Scientist

Scientists wonder how things work. They try to find out about the world around them. They do this by using scientific enquiry.

The diagram shows the important ideas about scientific enquiry.

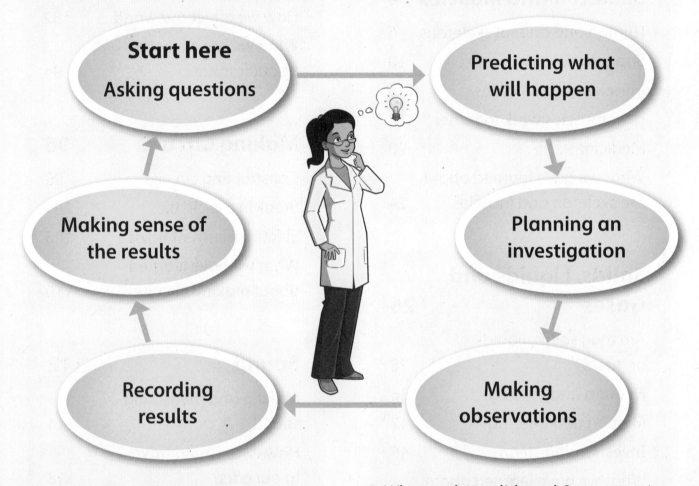

Start here
Asking questions

Predicting what will happen

Making sense of the results

Planning an investigation

Recording results

Making observations

An example investigation:

How can we stop some ice from melting?

- What makes solids melt?
- What will happen when I put the ice in a hot place?

Asking questions

Predicting what will happen

How can you ask questions?

Start your questions with words like 'which', 'what', 'do' and 'does'.

A prediction is when you say what you think will happen in your investigation.

A prediction is more than a guess.

Here is an example of a question and a prediction.

Question

Will putting the ice in a cold place slow down melting?

Prediction

Yes. Ice melts faster when it is hot.

Planning an investigation

When you plan an investigation think about how you will make it a fair test.

What will you keep the same?

What will you change?

Remember that the things you keep the same or change are called variables.

Making observations

You will measure time and test to find out whether putting ice in a cold place slows down melting. You will have to observe all of your ice samples.

To make your timing accurate you will use a clock.

Recording results

There are many ways to record results. A good way is to complete a table. A table keeps all of your results neat and tidy. It can help you to see patterns.

Making sense of the results

At the end of your investigation you must look at your table carefully.

You are comparing the conditions.

Were any of your results unusual?

Should you repeat your investigation to check how accurate your results are?

Are your results the same as other groups in your class?

Was your prediction correct?

Did your investigation make you think of any other questions?

1 Skeleton and Muscles

In this module you will:

- understand that humans and some animals have bony skeletons inside their bodies

- find out what skeletons are for, and that skeletons grow as you grow

- understand that animals with skeletons have muscles attached to the bones

- learn that muscles contract (shorten) to make a bone move and work together in pairs

- understand how we use medicines.

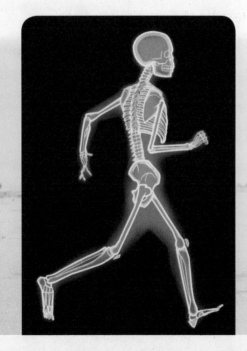

Amazing fact

Did you know that babies have more than 300 bones, but adults have only 206?

 What do you think happens to the extra 94 bones in a baby's skeleton?

Where do they go when the baby grows up?

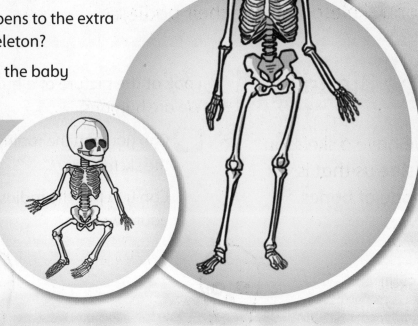

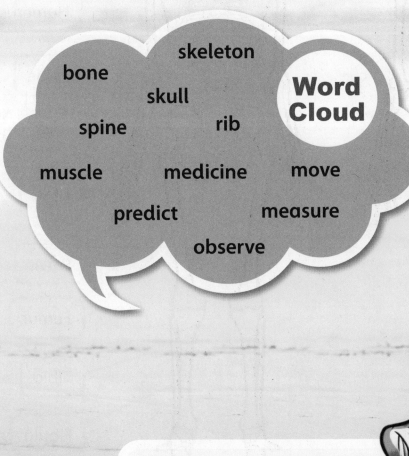

skeleton

bone

skull

spine   rib   **Word Cloud**

muscle   medicine   move

predict   measure

observe

Why do we need muscles?

Human and animal skeletons

Understand that humans and some animals have bony skeletons inside their bodies.

The Big Idea

We have a skeleton inside us that is made of bones.

Look at the picture of a human skeleton. We are humans.

Can you find the radius and ulna bones in the skeleton?

Can you feel the radius and ulna bones in your own arm?

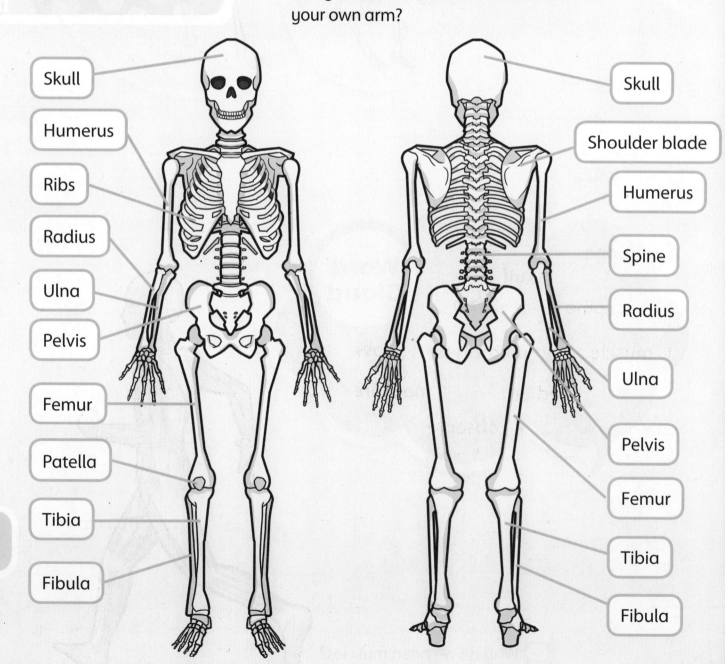

Skull
Humerus
Ribs
Radius
Ulna
Pelvis
Femur
Patella
Tibia
Fibula

Skull
Shoulder blade
Humerus
Spine
Radius
Ulna
Pelvis
Femur
Tibia
Fibula

Skeletons are made of bones that are white and hard. We all have a skeleton inside our body.

We can link the parts of the skeleton with the different parts of our body.

Starting at top of the skeleton and going down to the bottom:

- The **skull** is inside our head.

- The vertebrae go down our neck and our back. They are joined together to make the **spine**.

- The shoulder blades are inside our shoulders.

- The **ribs** are inside our chest.

- The humerus, ulna and radius are in our arms.

- The pelvis is around our hips.

- The femur is in the thigh.

- The patella is in the knee.

- The tibia and fibula are in our shins.

Which bones are in each part of the body? Draw lines to link the names of the bones to the correct part of the body.

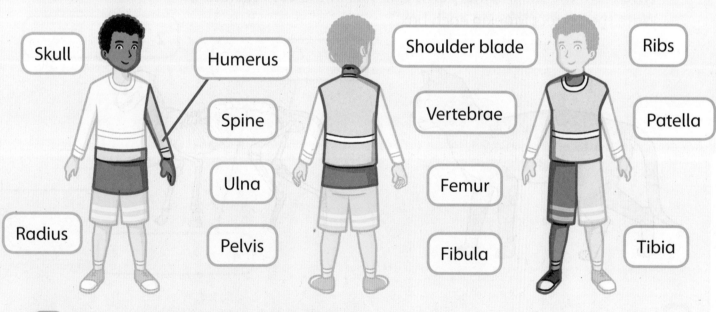

Skull Humerus Shoulder blade Ribs

Spine Vertebrae Patella

Ulna Femur

Radius Pelvis Fibula Tibia

Draw lines to match the two parts of these sentences.

Humans have a   ulna, radius and humerus bones inside them.

Skeletons are made of   femur, patella, tibia and fibula bones inside them.

Our legs have the   skeleton inside their body.

Our arms have the   skull inside it.

Our head has the   bone that is white and hard.

Human and animal skeletons

Understand that humans and some animals have bony skeletons inside their bodies.

1 s k u l l

2 r i b s

3 s p i n e

4 p e l v i s

The Big Idea

Human and animal skeletons have similar bones in them!

Look at the picture of a human skeleton. How much do you remember? Write the correct label in each box.

Look closely at the picture of a cat and its skeleton. Where are the skull, spine, pelvis and ribs in the cat's skeleton? Write the correct label in each box.

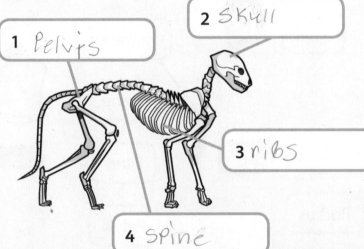

1 Pelvis

2 Skull

3 ribs

4 Spine

Look closely at the pictures of skeletons and the photos of animals. Which skeleton belongs to which animal?

Write the letter of each animal in the box next to each skeleton.

8

a
b
c
d
e
f

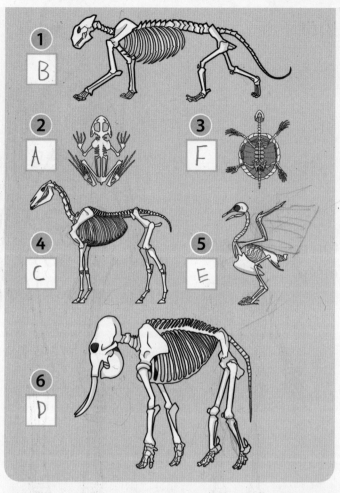

1 B

2 A

3 F

4 C

5 E

6 D

💬 Can you find the skull, spine, ribs and pelvis in each of the animal skeletons?

💬 What are the similarities and differences between the skeletons?

✏️ Now try these questions.

Which animal has ribs that are difficult to see?

Why do you think the frog has very long back legs?

How is the skeleton of the bird different from the other skeletons?

Name two bones that all these animals have.

and

✏️ Use the words in the word bank to complete the paragraph about animal skeletons.

Many animals have a ___skeleton___ inside their body. Animals have bones that are ___similar___ to the bones in a human skeleton. These bones include the ___skull___, the spine, the ribs and the pelvis. There are some ___diffrences___ between the bones of different animals. For example, birds have wings, ___elephants___ have tusks, and frogs have long legs.

Word Bank

~~differences~~ ~~similar~~ skull
elephants ~~skeleton~~

Now turn to page 24 to review and reflect on what you have learned.

Super skeletons

Find out what skeletons are for, and that skeletons grow as you grow.

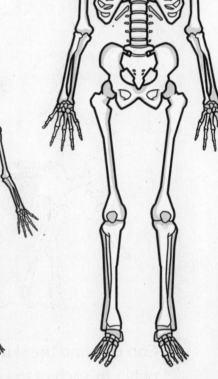

The Big Idea

Skeletons get bigger as we grow.

Look at the three skeletons. Which skeleton do you think is the closest to your own skeleton?

When we look at the skeletons from the baby to the adult, we notice that the skeleton in the adult is larger than the skeleton of a child. This suggests that skeletons grow. But are *all* the bones of a taller person bigger than the bones of a smaller person?

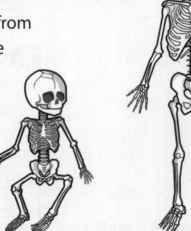

 Investigation: Do all bones grow at the same rate?

To answer this question, we need to find out if all the bones of a taller person are bigger than the bones of a smaller person. To do this, we need to **measure** some parts of the body.

1 Copy each table in your Investigation Notebook.

Height

Name of student	Height in cm
Aisha	127 cm

Hand length

Name of student	Hand length in cm
Ahmed	14 cm

Leg length

Name of student	Leg length in cm
Baserah	65 cm

Head circumference

Name of student	Head circumference in cm
Hunnain	43 cm

2 Measure the height, hand length, leg length and head circumference of each person in your group.

💬 Look at your results. How can we present the results so they are easier to read?

3 Use your results to create a bar chart in your Investigation Notebook.

If it is true that all bones grow at the same rate, the tallest person should have the biggest measurements for all the things you have measured.

✏️ Check your results to see if this is true or false. What can you conclude from your investigation? Are all the bones of a taller person bigger than the bones of a smaller person?

I conclude

✏️ Complete the paragraph about our investigation.

We investigated how skeletons grow by taking ___measurements___. We measured ___height___, hand length, leg length and head ___circumference___. These measurements gave us ___information___ about how our skeleton grows as we grow.

Word Bank

circumference	clues
~~measurements~~	height

Super skeletons

Find out what skeletons are for, and that skeletons grow as we grow.

The Big Idea

We need skeletons to support and protect our bodies.

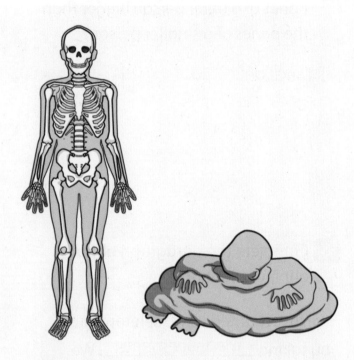

Skeletons do more than provide a framework. The functions of the skeleton are:

- to support our bodies

- to allow us to move

- to protect the organs inside our bodies, such as our heart, our lungs and our brain.

Let's look at the parts of the skeleton that protect organs.

Here are the skull and the brain. The skull is perfectly shaped to protect the brain.

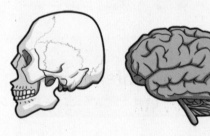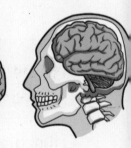

Here are the ribs. We call this structure the ribcage.

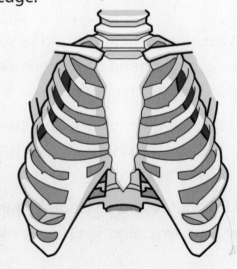

Q What would we look like if we didn't have a skeleton?

Why do we need a skeleton?

Skeletons provide a framework for the body of humans and other animals. The skeleton gives humans and animals their unique shape. When we look at each other, we know we are humans because of our shape.

It is the same when we look at a cat or bird or elephant. We know which animal it is because of its shape.

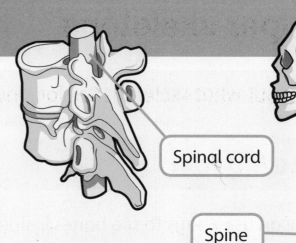

Spinal cord

Spine

 Investigation: Can you find your own ribcage?

Place your hands on each side of your ribcage with your middle fingers touching, Take a deep breath.

 What happens to your hands?

What happens to the ribcage?

It changes shape

The function of the ribcage is to protect some of the organs inside the body.

 Which organs does the ribcage protect? Look at the picture to find out.

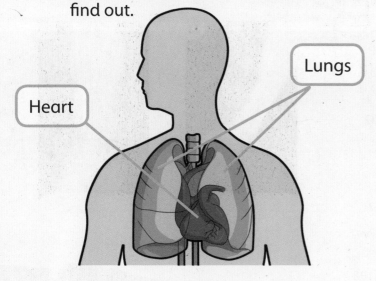

Heart

Lungs

H e a r t s and L u n g s

This is the spine.

The spine has two main functions:

- It supports our skull and helps us to stand straight and sit straight.

- It protects the spinal cord. (The function of the spinal cord is to carry messages to the brain.)

 Complete the sentences about the skeleton.

Humans have ___skeletons___ inside their bodies. The function of the skeleton is to allow __movement__, protect organs in the body and to support the body. The functions of the __spine__ are to protect the spinal cord, to support the skull and to help us to __stand__ and sit straight. The __skull__ protects the brain, and the __ribcage__ protects the heart and lungs.

Word Bank

<del>movement</del>   <del>ribcage</del>   <del>stand</del>
<del>skeletons</del>   <del>skull</del>   spine

Now turn to page 24 to review and reflect on what you have learned.

Super skeletons

Find out what skeletons are for, and that skeletons grow as we grow.

The Big Idea

We can use x-rays to see bones inside the body.

What do you remember about the skeleton?

 Complete the information about the parts of the skeleton. The first one has been done for you.

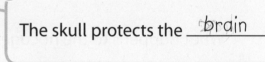

2 The skull protects the __brain__.

3 The ribcage protects the __heart__ and __lungs__.

1 The spine supports the __skull__ and helps us to stand straight and sit straight.

4 The spine protects the __spinal cord__.

The bones of the skeleton are very hard, but sometimes they can get broken. Breaks in bones are called fractures. We use x-rays to see fractures.

Look at these x-rays.

 Which bones do you think are broken? Write your answers in the boxes.

a

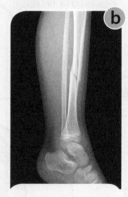

b

If we break a bone, it can be fixed. The bones are put back into place and held firmly in a cast until they get better.

Invertebrates

Not all creatures have skeletons.

Look closely at the picture. You can see that the crab does not have a spine. Creatures that do not have a spine are called <u>invertebrates</u>. Humans and other animals with a spine are called <u>vertebrates.</u>

 There are lots of different invertebrates. Draw lines to match the descriptions with the pictures.

1 **Jellyfish**: lives in the sea, has long tentacles

2 **Starfish**: lives in the sea, body has five parts

3 **Snail**: lives on land, has a shell on its back

4 **Spider**: makes webs, has eight legs

 a

 b

 c

 d

 Create a poster about vertebrates and invertebrates.

 Draw lines to match the two parts of these sentences.

Animals with a spine are called   jellyfish.

Animals without a spine are called   x-rays.

We can see broken bones using   invertebrates.

The invertebrate with tentacles is called a   vertebrates.

12/12/22

Muscles and skeletons

Understand that animals with skeletons have muscles attached to the bones.

The Big Idea

If humans and animals did not have muscles, they would not be able to move!

Look closely at the picture of the femur of a cheetah.

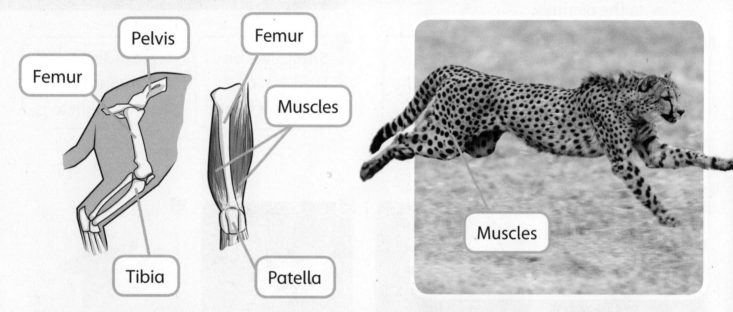

Femur
Pelvis
Femur
Muscles
Tibia
Patella
Muscles

What do you think muscles do?

Why do you think the cheetah needs strong muscles in its legs?

Do you think we need to have strong muscles in our legs?

Humans and animals need muscles so they can move. We can feel muscles all over our body.

Investigation: Let's try to find some muscles!

Put your left arm straight out and make a firm fist. Put your right hand on the muscle between your elbow and shoulder. Bring your fist towards your shoulder.

What happens when you bend your arm?

We can find muscles all over our body. The picture shows where the main muscles are. But what do they do?

💬 Which muscles do you think we use to do these activities?

Look at the picture to help you.

- Walking
- Talking
- Eating
- Running
- Breathing
- Picking up a pencil
- Swimming

How do muscles work?

We know that our skeleton is made of bones that are hard and white. Bones cannot move on their own. This is why we have muscles.

Think about our arms. We already know that we have three main bones in our arms: the humerus, the ulna and radius.

Where bones meet we have joints. For example, the elbow, the shoulder, the hip and the knee are joints. Muscles are attached to every bone. The muscles help us to **move** the bones. This is how we bend our arms and legs.

Humans and animal skeletons work in similar ways. All skeletons have muscles attached to the bones and they all have joints where bones meet up.

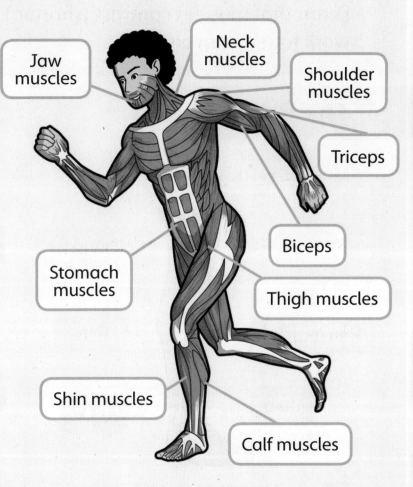

Jaw muscles
Neck muscles
Shoulder muscles
Triceps
Biceps
Stomach muscles
Thigh muscles
Shin muscles
Calf muscles

✏️ True or false? The first one has been done for you.

Muscles are attached to bones. (True) False

Bones can move on their own. True (False)

Elbows and knees are example of joints. (True) False

We use our jaw muscles for talking. (True) False

Now turn to page 25 to review and reflect on what you have learned.

How muscles work together

Learn that muscles contract (shorten) to make a bone move and work together in pairs.

The Big Idea

Muscles work in pairs!

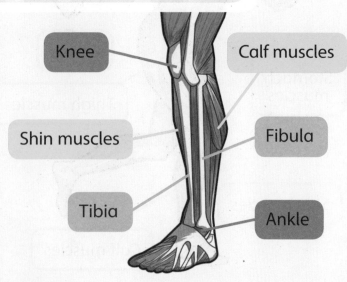

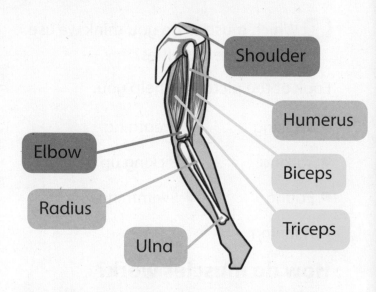

What happens to the biceps muscle when we bend our elbow?

Which muscles do we use when we point our toes?

Point your toes. Was your **prediction** correct? Yes or no?

Look at the picture of the muscles in your arms.

Which muscles do we use when we bend our elbow?

Our arm has two main muscles, the biceps and the triceps. They work together to move our arm. But how?

When our arm is straight, the biceps relaxes and the triceps contracts. Look at the picture. Notice the shape of the muscles.

The biceps is relaxed and is longer and thinner. The triceps is contracted and is shorter and fatter.

Bend your elbow. Was your prediction correct? Yes or no?

The function of the triceps is to pull the arm straight.

 Investigation: Find your triceps

Put your left arm straight down.
Use your right hand to find the triceps.
You can feel it at the back of your arm.

Look at the picture of the bent arm.
What do you notice about the muscles?

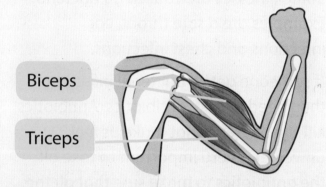

Biceps

Triceps

When the arm is bent, the muscles look different. The biceps is shorter and fatter and the triceps is longer and thinner.

This is because the biceps is now contracted and the triceps is relaxed. The function of the biceps is to pull the bones into the bent position.

 Investigation: What happens to the biceps when you bend your arm?

Put your left arm straight down. Use your right hand to find your biceps muscle at the front of your left arm. Slowly bend your elbow and feel what happens to the biceps muscle.

 Write down what you **observe**.

Muscles work in pairs all over our body.

Think about...
Can you think of another example where muscles work together in pairs?

?

 Write the missing words.

1 The opposite of relax is
   contract .

2 This muscle pulls the arm
   straight. _triceps_

3 We use this muscle to bend our
   arm at the elbow. _biceps_

4 Muscles work in _pairs_ .

5 When a muscle is contracted it is
   shorter and _fatter_ .

6 When a muscle is relaxed it is
   longer and _thinner_ .

7 The _humerus_ bone is in the
   upper arm.

Now turn to page 25 to review and reflect on what you have learned.

Medicines

Understand how we use medicines.

The Big Idea

We can take medicines to make us feel better when we are unwell.

💬 Have you ever been unwell?

When we feel unwell, sometimes we need to take medicines to help make us feel better.

What is a medicine?

A medicine works with the body to make us feel better. Some medicines can cure illness. Some medicines can make the symptoms better.

Coughs and colds

Have you ever had a cough? Cough medicines are used to help us feel better. Or we can use a natural treatment such as honey and lemon in hot water.

Have you ever had a cold? When we have a cold, we may feel hot and have a runny nose and a headache. There are many different kinds of cold medicines to help us feel better.

Antibiotics

Some illnesses are caused by bacteria. Examples are a sore throat, ear infections and chest infections.

If someone gets a bacterial infection, they need to take antibiotics. Antibiotics kill the bacteria that make us feel unwell. It is very important to take all the antibiotics to make sure that all the bacteria are killed.

💬 Have you ever taken antibiotics? When? Why?

Allergies

Some people have allergies, for example hay fever. People with allergies can take medicines called antihistamines.

Using medicines safely

It is very important to use medicines safely because if we take too much medicine, we can become unwell. Medicine labels have instructions about how much medicine to take and how often to take it.

Antihistamine

Dosage: Take 1 tablet once a day.

Warnings: Do not take if you are pregnant. May make you feel sleepy. Do not give to children under the age of 2.

Provides relief from:
- pet allergies
- skin allergies
- hay fever
- insect bites

Look at the medicine label. Write down:
- the type of medicine

  Ibuprohen (llama unicorn)

- what illnesses the medicine is used for

  Pain, fever, inflimation

- how much medicine to take

  ~~3~~ 2 tablets 3 times a day

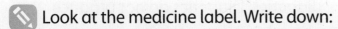

- how often to take the medicine

  3 times a day

- any warnings about the medicine.

  tummy ~~statch~~ scratch

Complete the sentences about medicines using the words from the word bank.

We sometimes take medicines when we feel ___*unwell*___. Medicines treat the ___*symptoms*___ and make us feel better. An example of a medicine is an antibiotic. Antibiotics are used to treat illnesses caused by ___*bacteria*___. It is very important to use medicines properly by reading the ___*instructions*___ on the label. Sometimes we do not need to use medicines because sometimes we can get better ___*without*___ them.

Word Bank

~~bacteria~~   ~~instructions~~   ~~unwell~~
~~without~~   ~~symptoms~~

Medicines

Understand how we use medicines.

The Big Idea

Some people need to take medicines for a long time.

💬 Look at the two children. One child has asthma. Which one do you think it is?

It is not always easy to know if someone has a long-term health problem because often they look very healthy. We will look at two different long-term health problems.

Asthma

Asthma is a long-term health problem that affects the lungs. We use our lungs when we breathe. The person with asthma may have problems breathing.

Asthma can be treated with:

- tablets to help prevent asthma attacks

- inhalers that help someone to breathe when they have an asthma attack. Inhalers spray medicines into the person's mouth and work quickly.

When someone has an asthma attack, they find it difficult to breathe. It is very important to keep them calm and make sure they take their medicine quickly. If you see someone having an asthma attack, tell a teacher. The teacher will look after the person until their breathing returns to normal.

Diabetes

People with diabetes are called diabetics. Diabetics do not produce enough insulin or the insulin may not work properly. Insulin helps our bodies to change a large sugar into a smaller sugar called glucose. Glucose gives us energy. Diabetics cannot control the amount of glucose produced without medicine or a special diet.

Many diabetics use a small blood test kit every day to check the level of glucose in their blood.

Some diabetics need regular injections of insulin to keep them healthy. The doctor shows them how to inject themselves.

Sometimes diabetes can be controlled by eating certain foods. The doctor will advise diabetics which foods to eat.

A diabetic has regular visits to their doctor to make sure their diabetes is under control.

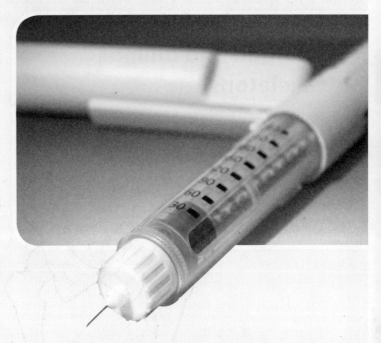

✏️ Do the crossword to see how much you remember about medicines.

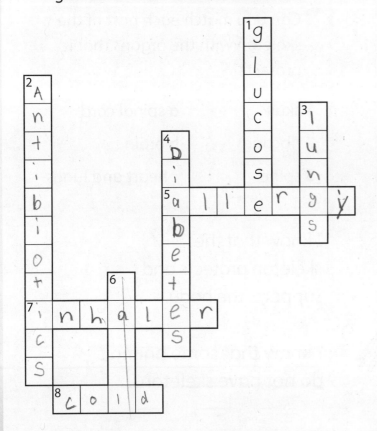

1 What substance do diabetics check for with a blood test kit?

2 Diseases caused by bacteria are treated with …

3 Asthma affects the small tubes leading to the …

4 This illness is treated by controlling insulin.

5 Hay fever is what type of health problem?

6 We can find instructions on how to use medicines on the …

7 The equipment used by asthmatics is an …

8 This illness can make us feel very hot with a runny nose.

23

Now turn to page 25 to review and reflect on what you have learned.

What we have learned about the skeleton and muscles

Human and animal skeletons

 Can you label the bones in the leg with the correct names?

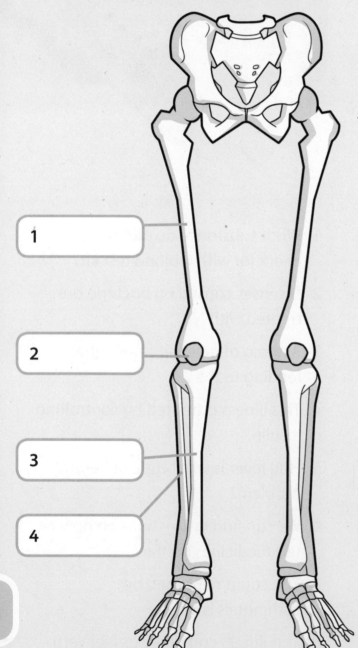

1

2

3

4

 Why have frogs developed long back legs?

I can name some of the bones in the human skeleton. ◯

I know some differences between the skeletons of different animals. ◯

Super skeletons

 Can you match each part of the skeleton with the organs that it protects?

1 skull	**a** spinal cord
2 ribcage	**b** brain
3 spine	**c** heart and lungs

I know that the skeleton protects and supports the body. ◯

I know that some animals do not have skeletons. ◯

Muscles and skeletons

 Why does our skeleton have muscles attached to the bones?

> I know why muscles are attached to bones. ◯

How muscles work together

 When one muscle contracts, what happens to the opposite muscle?

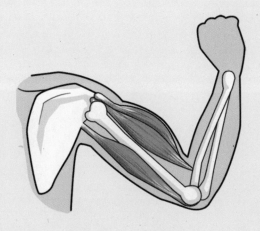

> I know why muscles work in pairs. ◯

> I know an example of a pair of muscles that work together. ◯

Medicines

 Before we take a medicine, what must we do?

People with diabetes cannot use glucose to produce energy. What medicine do they need?

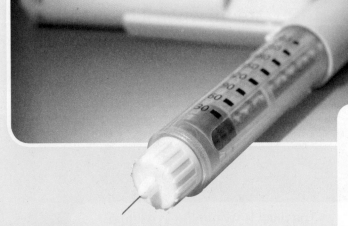

> I know that we can use medicines to make many common illnesses better. ◯

> I know that taking too much medicine can make us ill. ◯

2 Solids, Liquids and Gases

In this module you will:

- find out that matter can be solid, liquid or gas

- investigate how materials change when they are heated and cooled

- understand that melting is when a solid turns into a liquid and that melting is the reverse of freezing

- observe how water turns into steam when it is heated and that it turns back into water when it is cooled.

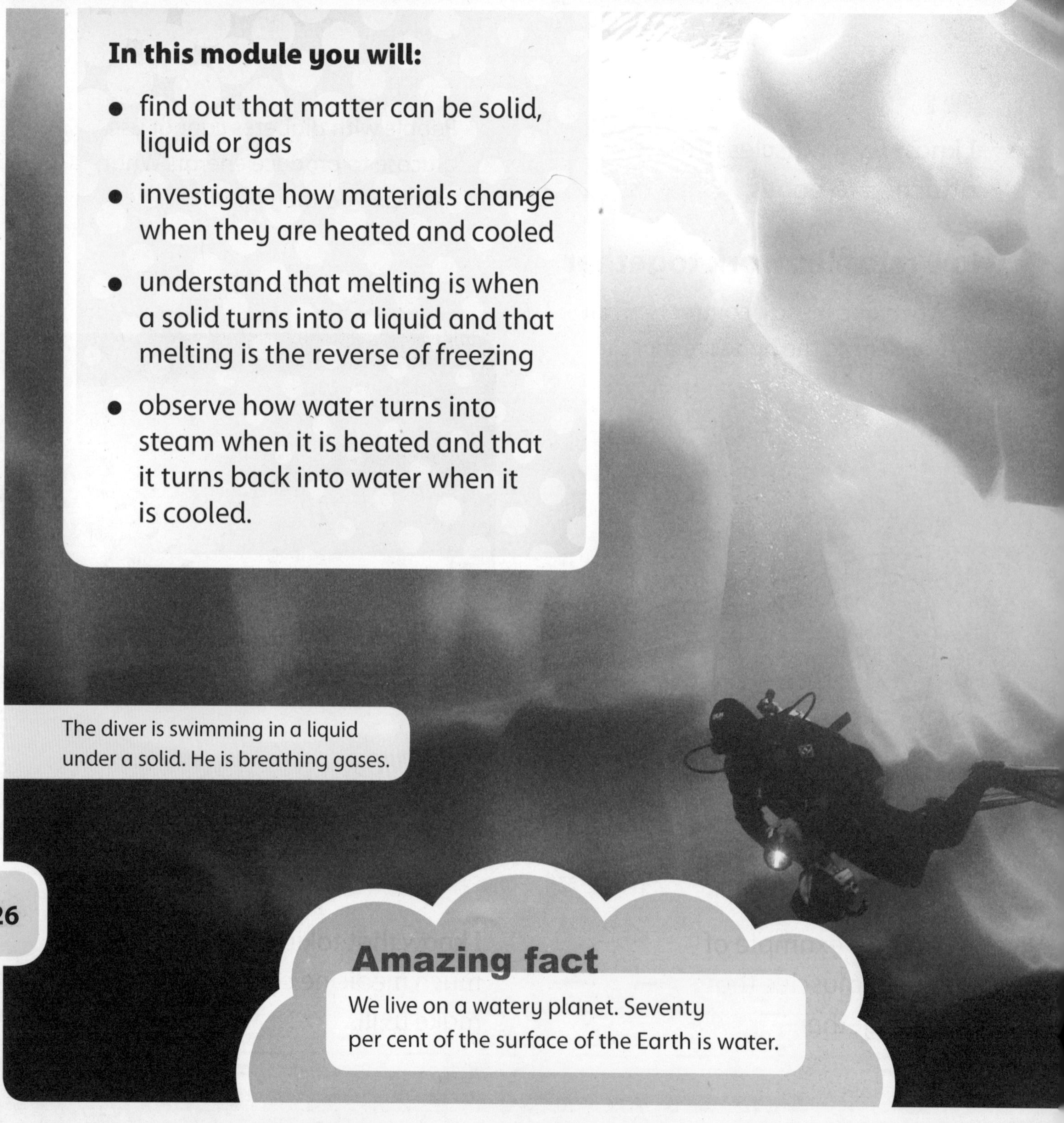

The diver is swimming in a liquid under a solid. He is breathing gases.

Amazing fact

We live on a watery planet. Seventy per cent of the surface of the Earth is water.

You have seen solids and liquids many times.

Discuss with your partner examples of solids and liquids you have seen today.

Think about foods and drinks, and things in the classroom.

Word Cloud

matter solid

gas melting

particle boiling

liquid record

freezing

Gases are not easy to see. There are gases in the air all around us but we cannot see them. They are transparent. Some gases do have colour. These are easier to see.

Look at the gas in the picture. How is it different from solids such as ice and wood?

Look at the picture of a kettle.

What is the white cloud?

Do you think the kettle is hot or cold? Why?

Chlorine gas is green.

27

Are they solids, liquids or gases?

Find out that matter can be solid, liquid or gas.

The Big Idea

We can find out if a substance is a solid, liquid or gas by looking at its characteristics.

1 Gas

2 Solid

3 liquid

Look at the picture.

✎ Label the picture by writing the words in the correct boxes.

Word Bank

solid   liquid   gases

The chair we sit on, the water we drink and the air we breathe are all made of different substances.

● A chair is solid.

● Water is liquid.

● Air is a mixture of gases.

✎ Look at the list of substances in the table.

Is each substance a solid, a liquid or a gas? Colour in the correct box.

Substance	Solid	Liquid	Gas
Blood		▨	
Carbon dioxide			✓
Milk		✓	
Copper	✓		
Stone	✓		
Oxygen			✓
Petrol		✓	✗
Paper	✓		
Wood	✓		
Plastic	✓		

💬 With your group make up some rules to help you identify the solids. Then do the same for the liquids and the gases.

Characteristics of substances

The characteristics of a substance describe the way it looks, feels and behaves. We use characteristics to show whether a substance is a solid, liquid or a gas.

We call solids, liquids and gases the three states of **matter**.

Characteristic	Solid	Liquid	Gas
Does it have a fixed **volume**?	Yes	Yes	No. It changes to fill the container.
Does it have a fixed **shape**?	Yes	No. It changes to fit the shape of the container.	No. It changes to fill all of the container's shape.
How **dense** is it?	Very dense	Dense	Not dense
How easy is it to **squash**?	Hard to squash	Hard to squash	Easy to squash
Does it **flow**?	No	Yes	Yes

💬 Look at the characteristics in the table. Did you include any of these in your own rules?

✋ Investigation: Identifying solids, liquids and gases

You are going to use the characteristics in the table to identify solids, liquids and gases.

Your teacher will give you some substances to identify.

1 Check each substance and group all the solids together.

2 Then group all the liquids together and then the gases.

3 **Record** your results in your Investigation Notebook.

✏️ Answer the questions about your investigation.

1 Which substances are solids?

2 Which substances are liquids?

3 Which substances are gases?

Are they solids, liquids or gases?

Find out that matter can be solid, liquid or gas.

The Big Idea

All matter is made of tiny particles that we cannot see.

Look at the photograph.

Which container is full of gases? How do you know?

Particles

If there was a microscope powerful enough, we could look at solids, liquids and gases in more detail. We would see the tiny **particles** and how they are arranged.

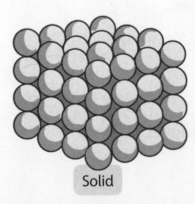

Solid

Liquid

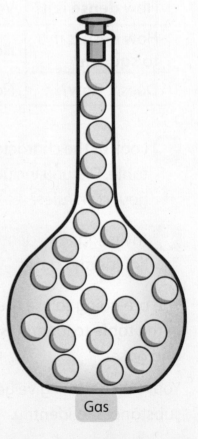

Gas

In the solid the particles are packed closely together. They do not move very much.

In the liquid the particles are not packed as closely together. They can move a small amount.

In the gas the particles are far apart. They can move quickly in every direction.

💬 Look at the drawings of the particles in a solid, a liquid and a gas. Discuss these questions with your partner.

- Why do you think solids are hard and have a fixed shape?
- Why do you think liquids have the same shape as the container and can be poured?
- Why do you think gases have no fixed shape or volume?

✏️ Look at the pictures. Under each picture draw how you think the particles are arranged.

| Oil | Air | Metal |

✏️ Can you answer these questions?

1 What are the three states of matter?

Liquid, Solid, gas

2 In which state are the particles closely packed together?

Solid

3 In which state are the particles free to move around in all directions?

gas

Are they solids, liquids or gases?

Find out that matter can be solid, liquid or gas.

The Big Idea

Powders behave like liquids but they are solids.

We know that liquids pour. We often pour water from one container (or from the tap) into another container.

 Which liquids can you see being poured in these pictures?

 Investigation: What is the volume and shape of liquids?

We are going to investigate what happens when we pour the same amount of water into different containers.

1 Select four different transparent containers.

2 Pour 100 millilitres (ml) of water into each container.

3 Look at the containers carefully.

- Do some containers look as if they contain more water?

- What shape is the water in each container?

- How easy was it to pour the water?

4 Record your observations in your Investigation Notebook.

Remember

A liquid changes shape to fit the shape of the container but it has its own volume.

When people sell liquids they sometimes put them into different shaped bottles. This is so they look attractive, but also so we think we are getting more for our money.

Think back to your investigation. You may have seen containers like these filled with water.

 Look at the containers and answer thequestions.

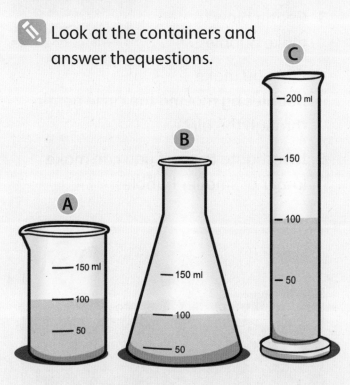

1 Which container looks as if it has the most water? B

2 Which container looks as if it has the smallest amount of water? C

3 Look at the measurements. What is the volume of water in each container?

A 100 ml   B 100 ml   C 100 ml

Investigation: Are powders liquids?

We can pour powders, so does this mean that they are liquids?

1 Take some sugar cubes. These are solids.

How do you know they are solids?

2 Pour the sugar lumps from one bowl to another.

They poured, but are they liquids?

3 Carefully crush the sugar cubes into very small pieces to make a powder.

4 Pour the powdered sugar from one bowl to another. It pours even better.

The powdered sugar pours, so is it a liquid?

5 Look at the powder with a hand lens or microscope. What do you see? Is the sugar still a solid?

6 Record your observations.

Remember
Powders pour but they are not liquids.

Are they solids, liquids or gases?

Find out that matter can be solid, liquid or gas.

The Big Idea

We can make and test gases.

Am I a solid, liquid or gas? Fill in the gaps by using the word 'solid', 'liquid' or 'gas'.

1 I am cold and white. I float on the sea when it is very cold. I am a ___Solid___.

2 I am in rain and you can pour me. You can see through me and drink me. I am a ___liquid___.

3 I am in the air. You breathe me in to stay alive. I am a ___gas___.

4 I am a powder used to make bread. You can pour me out of a bag. I am a ___Solid___.

5 I am used in cars and buses. You pour me from a pump in a garage. I am a ___liquid___.

6 I am hard and strong. I am used for making cars. I am a ___Solid___.

✋ Investigation: Making bubbles

1 Make a circle using thin wire. Leave enough wire to make a handle.

2 Dip your circle into soapy water.

3 Gently blow to make bubbles.

4 Can you make bubbles by moving the circle gently through the air?

5 Investigate to see if you can make larger or smaller bubbles.

6 Record your investigation in your Investigation Notebook.

Remember

Bubbles are made when gases are trapped inside a liquid.

Investigation: Making and testing gases

We are going to make a gas called carbon dioxide.

1 Carefully put three small spoons of baking soda into a balloon.

2 Pour vinegar into a small plastic bottle until it is about one third full.

3 Fit the balloon over the bottle opening. Be careful not to drop the baking soda into the bottle.

4 Hold up the balloon and slowly pour the baking soda into the vinegar.

What happens?

5 Record your investigation and your observations in your Investigation Notebook.

How do you know that a gas is being made?

> The balloon became larger.

6 Test your gas. Squeeze the balloon.

Is it hard or easy to squash the gas? Why?

> It is hard because it has a lot of gas

True or false? Circle the correct answer.

1 Solids have a fixed shape and volume.   True   *False*

2 Gases spread out to fill the whole container.   *True*   False

3 The particles in liquids are free to move around in any direction.   True   *False*

4 Powders can be poured so they must be liquids.   True   *False*

Now turn to page 50 to review and reflect on what you have learned.

Solids, Liquids and Gases

Heating and cooling

Investigate how materials change when they are heated and cooled.

The Big Idea

Substances change when they are heated.

💬 Discuss the photographs with a partner. What is the link between the two pictures?

Tell your partner one example of where you have seen heat in cooking.

Why do we heat some foods?

When we heat substances we add energy to them. The particles start to move faster and move further apart.

✏️ Draw what you think happens to the particles in the solid as it is heated.

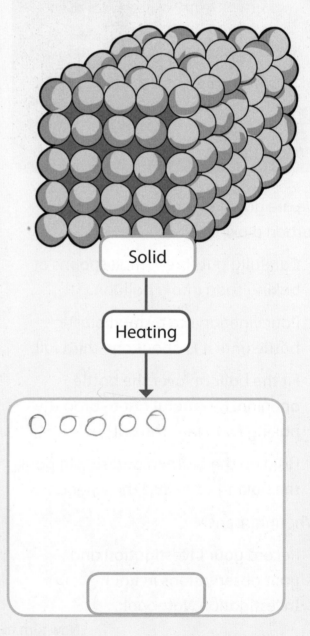

Solid

Heating

Investigation: How do materials change when we heat them?

We are going to heat some chocolate and observe what happens.

1 Put a heatproof bowl inside a metal pan.

2 Put your chocolate in the bowl.

3 Your teacher will pour hot water into the pan so that the bowl is standing in the hot water.

⚠ Do not touch the hot water or the bowl. Hot water is very dangerous.

4 Gently stir the chocolate for five minutes.

 Answer the questions about your investigation in your Investigation Notebook.

1 What did the chocolate look like before you put it in the bowl?

2 Was it a solid, liquid or a gas?

3 What did the chocolate look like after you heated it for five minutes?

4 Was it a solid, liquid or a gas?

Changes of state

When we heated solid chocolate it changed into liquid chocolate. This is an example of a change of state.

The change from a solid to a liquid is called **melting**.

 Look at the picture of an ice cream. Explain why this is an example of melting.

 Draw the particles in the ice cream when it is frozen.

Draw the particles in the ice cream after it has melted.

Heating and cooling

Investigate how materials change when they are heated and cooled.

The Big Idea

Liquids change state when they are heated.

💬 Discuss with your partner some examples of melting that you have seen.

Even metals can melt. A lot of heat is needed to make them change from solid metal to liquid metal.

Very hot liquid metal can be poured.

💬 Why is it useful to have melted metals. What can be made from them?

Water is a liquid. You have seen water being heated.

💬 Look at the picture. What do you see when water gets hotter and hotter?

🤚 Investigation: What happens when water boils?

Your teacher will show you what happens when water is heated.

⚠️ Do not get too close. Hot water is very dangerous.

✏️ Answer the questions.

1 What did the water look like before it was heated?

Liquid

2 What did you notice as the water became hotter and hotter?

It started boiling and turning into gas

3 What is the name of the white cloud that rises out of the water?

Steam

Evaporation

In warm weather or in a warm room, water particles can escape from the surface of water. The water slowly dries up. This is called evaporation. The liquid water changes to a gas called water vapour.

When we heat water, the heat gives the water extra energy. The water particles move faster and spread out. When the water is very hot the particles escape very quickly. You can see bubbles forming inside the water. When this happens the water is **boiling**. We call this very hot water vapour steam.

✎ Draw a line between each description and the correct word. One has been done for you.

1 On a warm day particles escape from the surface of water. This is called …

steam

2 When water vapour is very hot it is called …

evaporation

3 When water is heated bubbles form inside the water. This is called …

boiling

✎ Complete the diagram.

Draw the **three** arrangements of particles.

Write the **two** changes of state.

Write the missing state of matter.

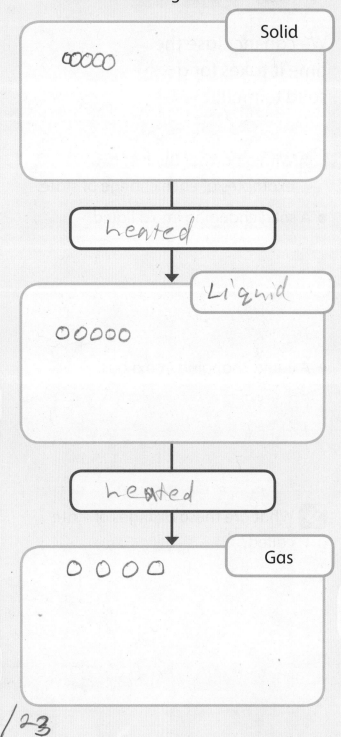

Solid

heated

Liquid

heated

Gas

1/11/23

Heating and cooling

Investigate how materials change when they are heated and cooled.

The Big Idea

We can increase the time it takes for a solid to melt.

With a partner think of two examples of each change of state.

- A solid changing into a liquid.

Ice into water
Chocolate bar into melted chocolate

- A liquid changing into a gas.

Water into steam
Liquid methene into gas.

What are these changes of state called?

melting and boiling

Global warming

Many scientists believe that global warming is melting the frozen parts of the Earth. The North and South Poles are a long way away, but the melting ice will cause problems for all of us.

How might the melting of the frozen parts of the Earth affect us?

The picture shows how much ice has melted at the North Pole since 1979.

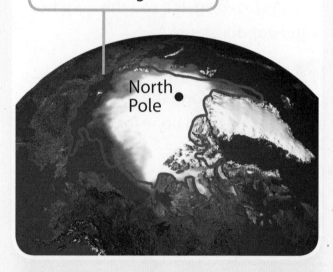

Ice boundary in 1979

North Pole

Investigation: How can we slow down melting?

We are going to try to keep ice cubes frozen for as long as possible.

💬 Why do ice cubes melt?

   If we do not have a freezer or a refrigerator, how can we slow down the melting?

1 Design a plan in your Investigation Notebook.

2 Try out your ideas and record the results.

💬 Share your results with the rest of the class. Discuss which ideas worked the best.

Before

During

Near the end

✏️ This is a page from a student's Investigation Notebook.

Can you fill in the words that have been washed out?

We have looked at different examples of **solids**, liquids and gases. When we **heat** a solid it changes into a **liquid**. This is a change of **state**. Heating the solid gives the particles more **energy**. They start to move more quickly.

We also heated liquids. These change into **gase**. When water changes to a gas in warm weather it is called **evaporation**. The gas is called water vapour. If we heat water until it is very hot, **steam** is made. This change of state is called **boiling**.

Word Bank

boiling	steam	energy
evaporation	heat	liquid
solids	gases	state

Solids, Liquids and Gases

Now turn to page 50 to review and reflect on what you have learned.

41

Melting and freezing

Understand that melting is when a solid turns into a liquid and that melting is the reverse of freezing.

The Big Idea

Almost all substances can melt and freeze.

Recycling metal

💬 Discuss the photographs with your partner.

What is happening to the metal in the middle picture?

Why does the furnace have to be so hot?

How many changes of state are there?

Melting and freezing

When a solid such as ice is heated it changes into a liquid. This process is called melting.

 Investigation: Recycling candles

We are going to investigate what happens when we heat a candle and then cool it again.

The candle is made of wax. When the wick is lit the candle burns. The wax slowly melts to a liquid. The heat melts the candle wax and it drips down the sides of the candle.

Some of this liquid evaporates to give a gas that helps the wick to burn better.

 Draw an arrow to show where the candle wax has melted and dripped down the side of the candle.

1 Press some objects into modelling clay to make moulds. Put your moulds on a piece of newspaper.

2 Your teacher will melt a candle for you and pour the wax into your mould. Leave the wax to cool.

> ⚠️ It is very dangerous to melt candles. Do not try to do this at home.
>
> Never put a flame near a melted candle. The liquid can burst into flames.

3 Carefully take the wax out of the moulds.

 Answer the questions about your investigation in your Investigation Notebook.

1 Is the wax a solid or a liquid?

2 What shape is the wax?

3 What changes of state have you seen in your investigation?

From liquid to solid

The change in state from liquid to a solid is called **freezing**.

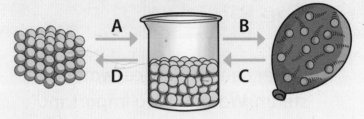

 Look at the picture and answer the questions.

1 Which arrow shows the change from a liquid to a solid? A, B, C or D? `D`

2 What is this process called?

> Melting

3 Which arrow shows the melting of a solid? `A`

4 What is process B called?

> Boiling

Melting is when a solid turns into a liquid. When a liquid turns back to a solid we call it freezing.

> ## Remember
> Freezing is the reverse of melting.

Solids, Liquids and Gases

43

Melting and freezing

Understand that melting is when a solid turns into a liquid and that melting is the reverse of freezing.

The Big Idea

Water can exist as ice, water and steam. Water is very important to us in all of its states.

💬 Work with a partner. Think of all of the times we have changed a liquid to a gas or a solid to a liquid.

Water is one of the most common substances on Earth. Approximately 70% of the surface is covered by water.

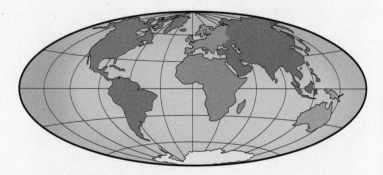

Water exists on Earth in all of its three states.

✏️ Write two examples of where we can find water in each of its states.

Solid ice

Glaciers, Ice cube

Liquid water

Ocean, Sea

Water vapour or steam

Steam, fog

✋ Investigation: At what temperature does water freeze and ice melt?

If you check in a book, it will tell you that ice melts at 0 °C. It will also tell you that water freezes at 0 °C. We are going to find out if both of these statements are true.

1 Set up the apparatus and carry out the investigations.

2 Record your investigation in your Investigation Notebook.

Finding melting point

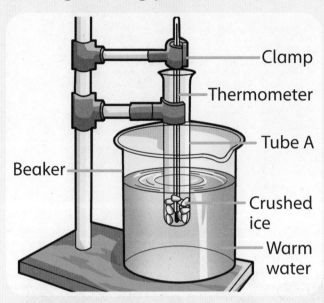

Clamp

Thermometer

Tube A

Beaker

Crushed ice

Warm water

Finding freezing point

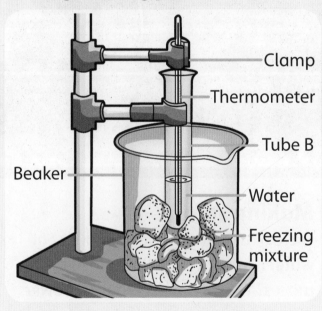

Clamp

Thermometer

Tube B

Beaker

Water

Freezing mixture

✏️ Answer the questions about your investigation in your Investigation Notebook.

1 What makes the water in tube B cool down?

2 What makes the ice in tube A warm up?

1/16/23

3 At what temperature did the water freeze? *0°C*

4 At what temperature did the ice melt? *0°C*

5 Did you prove that water freezes at 0°C and ice melts at 0°C? Yes or no? *Yes*

💬 Why is it more scientific to repeat the investigation a few times?

The more you do it the more acurasy you will have

✏️ Circle the correct word to complete each sentence.

1 When a solid changes to a liquid it is called evaporation / boiling / (melting).

2 When evaporation takes place a liquid changes to a solution / (gas) / solid.

3 The change from liquid to solid is called melting / filtering / (freezing).

✏️ True or false? Circle the correct answer.

1 Freezing is the reverse of melting.   (True) False

2 Water freezes at 0°C.   (True) False

3 The melting point of ice is 100°C.   True (False)

Now turn to page 51 to review and reflect on what you have learned.

Solids, Liquids and Gases

45

Investigating steam

Observe how water turns into steam when it is heated and that it turns back into water when it is cooled.

The Big Idea

Steam was once used as a means of power in industry and transport.

This train has been travelling in Jordan for many years. It is driven by steam. A big boiler heats up water and turns it to steam. This drives the wheels.

💬 Think of some other uses of steam engines. Discuss your ideas with a partner and write a list to share with the class.

⚠️ Steam is very hot. It can cause serious burns.

Making steam

Have you noticed that when we heat water in a pot with a lid on the lid moves? The lid is moved by the power of steam. Steam particles move very quickly and spread out. They collide with the lid and make it move.

If we can make the steam travel down a tube, we can make it do work.

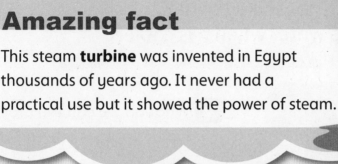

Amazing fact

This steam **turbine** was invented in Egypt thousands of years ago. It never had a practical use but it showed the power of steam.

Discuss the steam turbine with a partner. Study the diagram and try to describe how it works. Explain your ideas to the class.

Investigation: How hot is steam?

Your teacher will demonstrate how to measure the temperature of steam. Your teacher will boil the water and hold a thermometer in the steam.

Water boils at 100 °C.

What temperature is the steam?

Why do you think the temperature of the steam is not exactly 100 °C?

Superheated steam

If steam is trapped in a container it is possible to keep heating the steam up. The pressure builds up as the steam particles move faster and faster and get more and more energy. Steam for engines is usually 500 °C. This is called superheated steam.

What is superheated steam?

These pipes are full of superheated steam.

Think about...

What do you think might happen to a container full of steam if you keep heating it for a long time?

Investigating steam

Observe how water turns into steam when it is heated and that it turns back into water when it is cooled.

The Big Idea

When steam cools down it turns back to water.

When the water in a kettle boils it makes steam. Making electricity can also make a lot of steam.

✏️ What do we call the process when water bubbles and makes steam?

What happens to the steam?

When the steam cools down it changes back to water. This happens when the steam touches a cool surface.

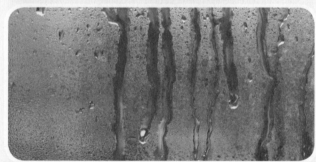

💬 Where do you think the steam goes?

When water is warmed it changes to water vapour. This is a change of state from liquid to gas.

✏️ What is the change of state from a liquid to a gas called?

💬 Where have you seen steam changing back to water? Discuss your ideas with a partner.

When a gas changes to a liquid it is called condensation. The steam condenses to water. This often happens in bathrooms and kitchens.

If the water is heated to a higher temperature it bubbles. Very hot water vapour is made. This is called steam.

48

Changes of state

We have now looked at all of the changes of state.
We can show all the changes of state in one diagram.

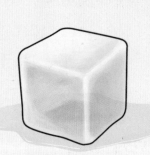 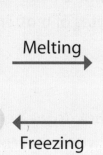 Melting →  Evaporation →

← Freezing ← Condensation

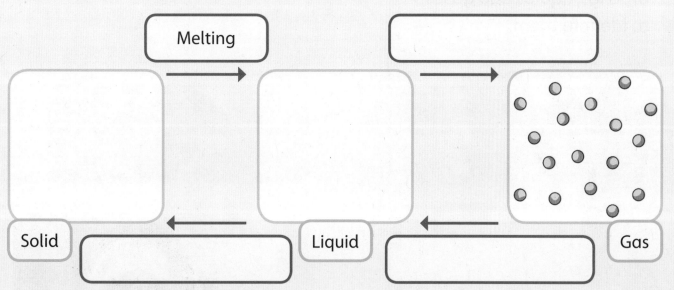

Complete the diagram showing changes of state. Draw the arrangements of particles in the big boxes. Write the labels in the smaller boxes.

Melting →

Solid   Liquid   Gas

Look at these words.

liquid	solid	100 °C
0 °C	condensation	
evaporation		
gas		
melting	freezing	

Draw a circle around the changes of states.

Tick the melting point of water.

Draw a star next to the boiling point of water.

Draw a box around the states of matter.

Now turn to page 51 to review and reflect on what you have learned.

What we have learned about solids, liquids and gases

Are they solids, liquids or gases?

✏️ Explain why gases can fill large spaces or be squashed into tiny spaces.

I can use the characteristics of solids, liquids and gases to identify them. ⭕

I know how the particles are arranged in solids, liquids and gases. ⭕

Heating and cooling

✏️ What are the names of the three states of matter?

✏️ What do we need to change ice to water and to change water to steam?

I understand what happens to the particles when a substance changes state. ⭕

Melting and freezing

To change a solid to a liquid we put heat into the substance.

 What happens to the heat in the liquid when we freeze it back to a solid?

I understand that melting is the reverse of freezing.

Investigating steam

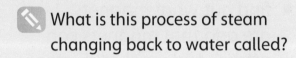 How do you change steam back to liquid water?

What is this process of steam changing back to water called?

Superheated steam reaches temperatures of 500 °C. Why is this better than steam from the kettle for driving steam engines?

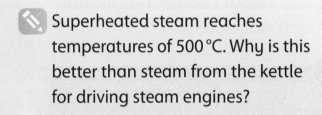

I understand that steam contains heat energy and that can be used to drive machines.

I can draw a diagram showing all the changes of state.

3 How Magnets Work

In this module you will:

- find out what magnets are and that magnets can attract and repel each other

- find out that magnets attract some metals, but not others.

Amazing fact

About a million years ago the Earth's magnetic poles were reversed. The North Pole was in the South and the South Pole was in the North. This may happen again. Do not worry – it won't happen for another million years.

The Earth's magnetic field protects us from radiation from the Sun. When solar rays hit the magnetic field it causes magnetic storms like the Northern lights.

Word Cloud

magnet
force pole repel
attract bar magnet
North South
iron steel investigate
test

Maglev trains use magnetic force to lift the train slightly from the track. A Maglev train in Japan travels at about 500 kilometres per hour.

Investigation: Make an electromagnet

Electromagnets are strong magnets that use electricity. We can make our own electromagnet.

Battery

Terminals

Plastic-coated copper wire

Steel nail or screw

⚠ The terminals of the battery can become warm, so take care.

May the force be with you!

Find out what magnets are and that magnets can attract or repel each other.

The Big Idea

Magnetism is an invisible force that can lift very heavy weights.

The electromagnet in the photograph could be a million times stronger than the one you made.

Where do people use very strong electromagnets to lift things?

How do magnets work?

Magnetism is an invisible **force** of attraction between some metals. The magnetic force comes from the billions of tiny particles called atoms that make up the metal. Each atom is like a very tiny **magnet**.

Magnets have two ends called **poles**. One end is the **North** pole and the other end is the **South** pole. In non-magnetic metals all the tiny magnetic atoms point in different directions.

In a magnet all the North ends of the tiny magnetic atoms point North and all the South ends point South. When the atoms are lined up, an invisible magnetic force appears.

Non-magnet

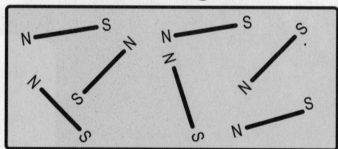

Magnet

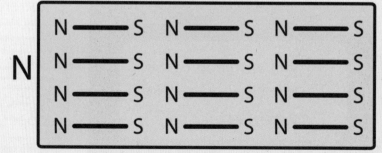

54

Why does the Earth have a North Pole and a South Pole?

At the centre of the Earth there is a liquid core. The core is made of molten metals that are magnetic. It creates a massive magnetic pulling force at each end. This force wraps around the Earth's surface and creates the Earth's magnetic field with a North end and a South end.

The North end of a **bar magnet** is attracted to the North Pole. The South end of a bar magnet is attracted to the South Pole.

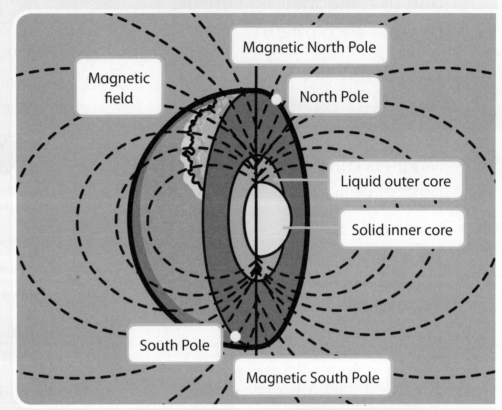

Magnetic North Pole

Magnetic field

North Pole

Liquid outer core

Solid inner core

South Pole

Magnetic South Pole

What is a compass?

A compass is a magnet that can turn freely so it always lies in a North to South direction.

💬 When is it helpful to have a compass?

How does a compass help us find our way?

✏️ Draw lines to match the beginnings and endings of the sentences about magnets.

1 Inside a metal magnet there are   a North pole and a South pole.

2 Each atom has   to make a compass.

3 When all the atoms point in the   millions of tiny particles called atoms.
   same direction

4 The North pole of a magnet is   the metal becomes magnetic.

5 We can use this   attracted to the Earth's North Pole.

May the force be with you!

Find out what magnets are and that magnets can attract or repel each other.

The Big Idea

Magnets can either push or pull.

We are going to **investigate** what happens when we bring two magnets together.

💬 What do you predict will happen?

✋ Investigation: How do magnets react together?

You will need two bar magnets.

In your Investigation Notebook, for each investigation:

- write your prediction
- copy the picture and label the North end N and the South end S
- slowly bring the two magnets towards each other
- observe if the magnets pull together (**attract**) or push apart (**repel**)
- write your conclusions for each investigation.

Circle the correct word.

1 North end facing South end

✏️ The magnets moved apart / together.

Opposite ends of the magnets attract / repel each other.

2 North end facing North end

✏️ The magnets moved apart / together.

The North ends of the magnets attract / repel each other.

3 South end facing South end

✏️ The magnets moved apart / together.

The South ends of the magnets attract / repel each other.

4 North end above North end and South end above South end

✏️ The magnets pulled together / pushed apart.

Like ends of the magnets attract / repel each other.

5 North end above South end and South end above North end

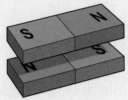

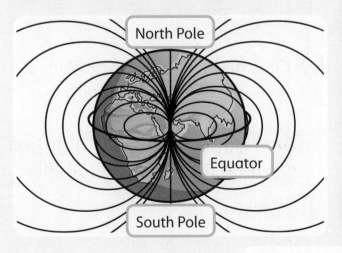 The **magnets** pulled together / pushed apart.

Opposite ends of the magnets attract / repel each other.

The law of magnetism is:

'Opposite poles attract and like poles repel.'

💬 Did your investigations show you this?

The magnetic forces on Earth come out at the North Pole and the South Pole. These two forces are attracted to each other. They bend around the Earth's surface to meet in the middle.

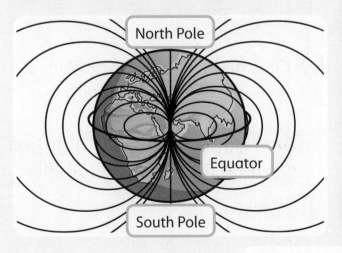
North Pole
Equator
South Pole

Think about...
Where do you think the magnetic field is the strongest? At the Poles or at the Equator?

✏️ How much have you learned about magnets? Write the sentences in the correct order.

1 particles   Inside   there are   metals   millions of

2 and a   North end   has a   Each particle   South end

3 North–South   In a   all the particles   are lined up   magnetic metal

4 attract   magnet   of a   poles   Opposite

5 poles   of   magnet   a   repel   Like

How Magnets Work

Now turn to page 62 to review and reflect on what you have learned.

Which things are magnetic?

Find out that which things are magnetic and how we use magnets.

The Big Idea

You can use a magnet to find out which materials are magnetic.

We are going to use a magnet to **test** objects made of different metals and other materials.

💬 Which of these things do you think is magnetic?

Chair leg (steel / aluminium)

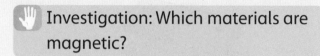
Window (glass)

Table (wood)

Pan (cast iron)

Bowl (plastic)

Drink can (aluminium)

Water pipe (copper)

Food can (steel)

✋ Investigation: Which materials are magnetic?

1 Copy the table in your Investigation Notebook. Write the objects you test in the table.

2 Use a magnet to test the objects. If the magnet is attracted, the material is magnetic. Tick the box Yes or No.

⚠ Do not test electronic objects such as mobile phones, interactive whiteboards or computers. The magnet will damage them.

Object	Material	Magnetic? Yes	No
Chair leg	Steel / aluminium		

✏ Copy and complete the table in your Investigation Notebook.

Objects that are magnetic	Objects that are non-magnetic

Complete the sentences.

We found out that all the magnetic objects are metals, but not all the objects made of metal are magnetic. Only the objects made of ___iron___ and _____ are magnetic.

The objects made of _____ _____ _____ are non-magnetic.

Word Bank

~~iron~~	brass	copper
wood	paper	steel

Magnetic metals

The Earth's core is magnetic. It is made of three metals: **iron**, nickel and cobalt.

Iron, cobalt and nickel are the only magnetic metals.

Iron is very common on the Earth's surface and nickel is quite common, but cobalt is extremely rare.

Many metal objects are made of steel. Why is steel magnetic?

Steel is a mixture of metals. It is mainly iron with some nickel. Metals that are mixed together are called alloys. Steel is an alloy of iron.

Not all steel objects are magnets.

They only behave like a magnet when they are magnetised.

To make a magnet all the tiny Norths and Souths in the atoms need to be lined up.

 **Investigation: Make a magnet**

1 Hold the nail flat and stroke the magnet along the nail.

2 Make sure that you only stroke in one direction.

3 Repeat this 50 to100 times.

4 Now test the magnet you have made. How many paperclips does it pick up?

 Which of these metals are magnetic? Draw a circle around the magnetic metals.

Copper   Iron   Lead

Tin   Steel   Brass

Nickel   Aluminium

Gold   Silver

Which things are magnetic?

Find out which things are magnetic and how we use magnets.

The Big Idea

Magnetic force is used in hundreds of things that we use every day.

Magnets can find the Earth's North Pole and South Pole.

How does the compass work? What is the 'needle' made of?

Investigation: Make a compass

Ask your teacher to help you make a compass.

Magnets attract metals that contain iron and not steel

Recycling centres use magnets to separate aluminium and steel cans. The magnet attracts the steel cans and leaves the aluminium cans behind.

We use fridge magnets to hold messages on the fridge door.

Investigation: Make a fridge magnet

Ask your teacher to help you make a fridge magnet.

Electricity can make very strong magnets

The magnet on the end of the crane is powered by electricity. This makes the magnet very strong. When the crane driver switches off the electricity the metal loses its magnetism, so the driver can put the scrap steel down.

Electromagnets are used in door locks. When the electricity is switched off, we can open the door.

Investigation: Make a stronger electromagnet

Do you remember the electromagnet you made? Can you make a stronger one?

Magnets can push as well as pull

Do you remember that the like poles of a magnet repel each other? This is how the Maglev train works. The magnets in the train and the magnets in the steel track push each other away. This lifts the train so it moves much more easily.

Magnets are used in many devices

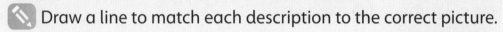 Draw a line to match each description to the correct picture.

1 Inside this building is an extremely big magnet that helps to make electricity.

2 Inside this device electricity makes a tiny magnet vibrate (shake). The vibrations create sound.

3 Inside this tool is an electric motor. Motors use electricity to make magnets that spin around very quickly.

4 This machine uses an electric motor to spin the washing.

 In three of the pictures electricity is used to make magnetism. Which picture is the odd one out?

Electric drill

Power station

Washing machine

Speakers

Think about...

There are many electric motors in a modern house, from the vacuum cleaner to the ceiling fan.

Find out where else magnets are used in the home.

The Maglev train and the crane both use magnetic force.

Does the magnetic force attract or repel? Write the correct word.

Maglev train

Crane

Now turn to page 63 to review and reflect on what you have learned.

What we have learned about how magnets work

May the force be with you!

 What causes the Earth's magnetic field?

 Look at the diagram of the atoms in a non-magnet. Can you draw how the atoms are arranged in a magnet?

Non-magnet

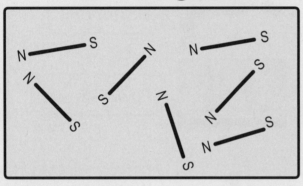

Magnet

N **S**

I understand how a magnet works. ◯

I understand the law of magnetism.

I understand how the North and South poles of a magnet react to each other. ◯

Which things are magnetic?

✏️ There are only three magnetic metals. What are they?

[ ]

✏️ Why is steel magnetic?

[ ]

✏️ Why does a magnet make a good compass?

[ ]

✏️ Can you name at least four different things that use a magnet?

[ ]

I know which metals are magnetic. ◯

I can make a magnet. ◯

I know that magnets are used in lots of ways in the home and in industry. ◯

4 Habitats

In this module you will:

- discover that different animals live in different habitats and how animals are adapted to their habitat

- learn how to use identification keys

- find out how human activity affects the environment.

1

2

3

Word Cloud

pollution

habitat   animal   plant

natural disaster   identify

pooter   equipment   volcano

bar chart   environment

Look closely at the map of the world and the photographs. The photographs show different animals living in different habitats.

💬 What do you think the word 'habitat' means?

✏️ Which habitats can you see in the photographs? Write the name of each habitat in the correct box.

Word Bank

Deserts   Grasslands   Oceans
Polar regions   Tropical forests

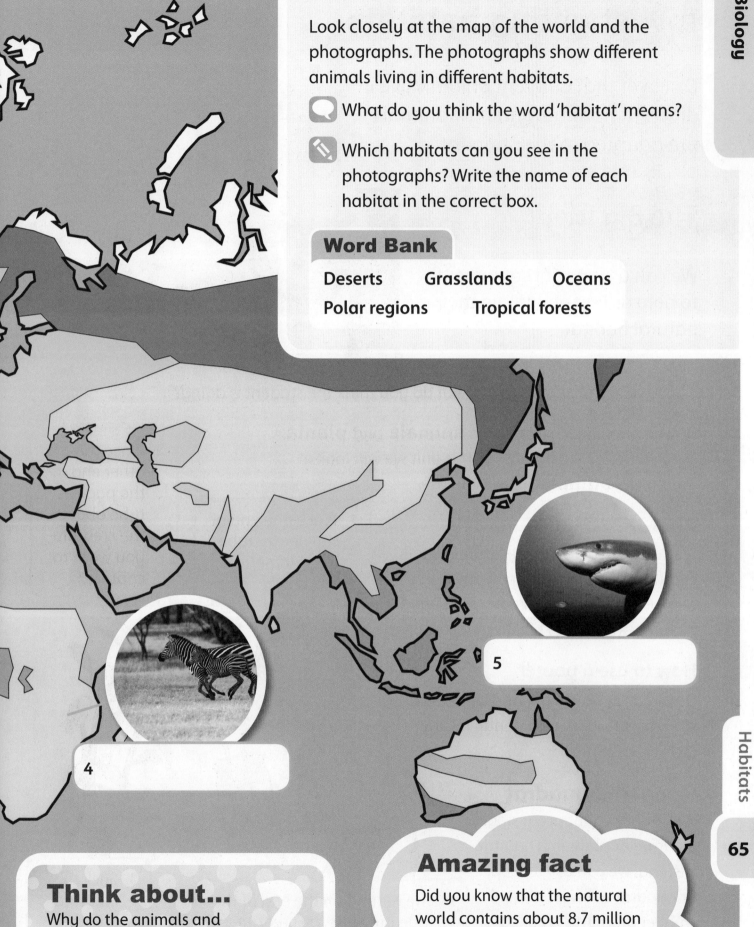

5

4

Think about...
Why do the animals and plants live in these habitats?

Amazing fact
Did you know that the natural world contains about 8.7 million species of plants and animals?

Investigating habitats

Discover that different animals live in different habitats and how animals are adapted to their habitat.

The Big Idea

We can use special equipment to help us find animals in their natural habitat.

💬 Look at the photograph. What do you think the student is doing?

How can we investigate which **animals** and **plants** live in different **habitats**? In this unit we will look at the special **equipment** we can use.

Pooter

We can use a **pooter** to capture small creatures. We use an identification key to help us **identify** the creatures. It has pictures and names of different creatures.

How to use a pooter

Once you have caught and identified the creature, you must gently return it to its habitat.

Hoop and quadrat

We cannot count every single creature or plant in a habitat. Instead, we can use a hoop or quadrat to investigate a smaller area. This is called sampling the habitat.

1 Place one end of the pooter tube in your mouth.

2 Place the other end of the pooter tube above the creature you want to capture.

3 Suck the creature into the pooter.

4 The netting on the end of the tube prevents the creature going into your mouth!

When scientists investigate a habitat, they get information from different places in the habitat. This gives them a good idea of the plants and creatures that live in the whole habitat.

How to use a hoop or quadrat

1 Choose the place carefully. The place you choose needs to show a good example of what we can find in the habitat.

3 Identify and count all the creatures or plants you find inside the hoop or quadrat.

2 Place the hoop or quadrat on the ground in the habitat you are investigating.

Sweep net

We can use a sweep net in three different ways to capture small creatures.

 Match the photographs of using a sweep net with the creatures you can catch.

Remember
You must identify the creatures quickly and return them gently to their **environment**.

1 Sweep the air

2 Sweep low-growing plants

3 Sweep a pond or stream

a Creatures that live in water

b Creatures that live in plants

c Flying creatures

 Create a poster about investigating habitats.

Investigating habitats

Discover that different animals live in different habitats and how animals are adapted to their habitat.

The Big Idea

We can investigate a habitat in the local environment!

We are going to plan and conduct an investigation into small creatures in the local environment.

 What types of creatures will you be looking for?

Investigation: What creatures live in local habitats?

Planning the investigation

Investigation title

Decide on a title for your investigation.

Equipment

Write a list of the **equipment** you will need.

Locations

Decide where you will carry out your investigation. What will you look for when you get there? Here are some examples of creatures you might see.

Scarab beetle

Ant

Butterfly

Your teacher will give you a map of the local habitats you will be investigating.

 Look at the map. Discuss in your group the route you will take and the locations you will investigate.

 Use your Investigation Notebook to write your investigation plan. Include:

- title of investigation
- equipment
- locations
- predictions of what you will find.

Carrying out the investigation

What is the best equipment to use at each location?

- Sometimes you will use the pooter and the hand lens. Remember to count how many creatures you see.

- Sometimes you will use the quadrat or hoop. Count how many different creatures you find inside the quadrat or hoop, and count how many of each type of creature you see.

Look carefully for creatures in cracks, and underneath stones or leaf piles.

⚠ Do not touch the creatures with your hands. Some creatures bite!

Remember
Always return the creatures to their habitat when you have identified them. If you see a creature you are unsure about, ask your teacher what it is.

 Use your Investigation Notebook to record the creatures that you find. Copy and complete the table.

Location	Equipment used	Creatures	Number of creatures
Footpath	Quadrat Hand lens	Ants	50
		Beetles	5

Investigating habitats

Discover that different animals live in different habitats and how animals are adapted to their habitat.

The Big Idea

We can present data from habitat studies in bar charts and graphs.

How can you present the data you collected in your investigation of local habitats?

Now we have completed our investigation. We have collected information about creatures in the local habitats. We have recorded the results in a table.

What happens next?

Interpreting data

The word 'data' means information. To interpret and understand the data we need to ask the question: 'What is this data telling us?'

To help us answer this question, we can present the data in a **bar chart**. This makes the information easier to read and interpret.

Here is the table that one group used to record the number of creatures they found.

Location	Creatures	Number of creatures
Site 1: Footpath	Ants	50
	Caterpillars	0
	Beetles	5
Site 2: Playing field	Ants	25
	Caterpillars	1
	Beetles	8
Site 3: Flowerbed	Ants	10
	Caterpillars	30
	Beetles	10

They used their data to draw this bar chart.

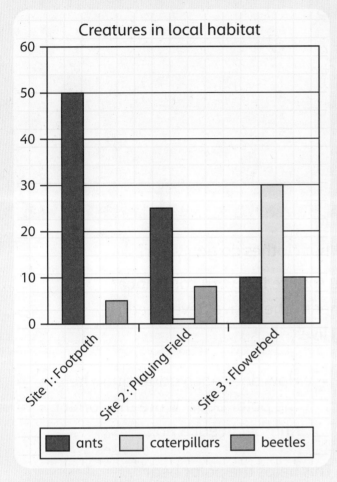

Creatures in local habitat

What does this bar chart tell us?

The group used their bar chart to interpret their data. This is what they wrote. Look closely at the bar chart and fill in the gaps in the group's interpretation.

There are more ants at site __1__ than at sites __2__ and __3__.

Ants seem to prefer to live on the ___footpath___ and ___playing field___.

They don't seem to like to live on the ___flowerbed___.

There are more caterpillars at site _____ than at sites _____ and _____.

Caterpillars seem to prefer to live on the _____. They do not seem to like living on the _____ or the _____.

There are more _____ than _____ in the habitats we investigated. Beetles seem to like all three sites. We found beetles on the footpath, on the playing field and in the flowerbed.

We did not find as many _____ as _____. For example, at site 1 we found _____ ants but only _____ beetles.

 Now interpret your own data.

Use the data you collected with your group to create a bar chart in your Investigation Notebook. Use the heading 'Group data'.

 Use your bar chart to help you write some sentences to interpret your data.

Habitats

71

Investigating habitats

Discover that different animals live in different habitats and how animals are adapted to their habitat.

The Big Idea

Many animals are adapted to live in their environment.

💬 Imagine it is a very cold and rainy day. What clothes do you wear?

Animals do not wear clothes, so how can they survive in a very cold place? Animals have adapted to their habitat. This means they have developed features that help them to survive where they live.

Polar bears

Polar bears have adapted to live in the very cold polar regions of the Arctic.

✏️ How does each adaptation help the polar bear? Write the correct number in each box.

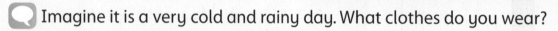

1 Excellent sense of smell	2 Thick white fur

3 Strong legs	4 Thick, rough paws

This helps the polar bear …

to run fast and to swim fast in icy water. ☐

to find the animals that it hunts. ☐

to walk on slippery ice. ☐

to hide in the snow and to keep warm. ☐

Penguins

Some penguins live in the polar regions of Antarctica. Penguins are birds.

Penguins need to keep warm. They also need to swim very well to catch fish.

 How does each adaptation help the penguin? Tick the correct box in the table.

Look closely at the picture of penguins to help you decide.

Adaptation	This helps the penguin ...	
	to swim	to keep warm
Shape of the penguin's body		
Lots of body fat (blubber)		
Shape of the flippers (wings)		
Shape of the feet		
Feathers close together		
Shape of the tail		
Strong chest muscles		
Dark feathers on their backs		
Feathers on the baby penguin		

 Create a poster showing the adaptations of a different creature that lives in a cold habitat.

Investigating habitats

Discover that different animals live in different habitats and how animals are adapted to their habitat.

The Big Idea

Some animals are adapted to live in the desert.

💬 Imagine a very hot day. How do you stay cool?

We will be looking at how animals are adapted to live in very hot deserts.

Fennec fox

The fennec fox is the smallest fox in the world. It can survive in the desert because it is adapted to its habitat. The Sahara Desert is very hot during the day but can be quite cold at night.

- The fennec fox has huge ears. These help to cool the fox down when it is very hot.

- It has thick fur that keeps it warm during the cold nights. The fur also protects the fox from the sun's heat during the day.

- The fennec fox is usually active during the night when it is cooler.

💬 The fennec fox lives in underground dens and has hairy paws. Why do you think it has hairy paws?

Camel

Look at the picture of the camel.

 How does each adaptation help the camel? Write the number of each adaptation in the correct box.

1 A hump on the camel's back

2 Long, thick eyelashes

3 Nostrils that can open and close

4 A tough, leathery mouth

6 Webbed feet with two toes

5 Tough, leathery knee pads

These help the camel to kneel on the hot sand. ☐

These protect the camel's eyes from the sand and the Sun. ☐

These prevent the camel from breathing in sand. ☐

These prevent the camel from sinking in the sand. ☐

This stores fat so the camel can go without food for a long time. ☐

This helps the camel to chew tough, thorny plants. ☐

 Correct the mistakes in these sentences. Cross out the word that is wrong and write in the correct word.

Polar bears have adaptations to help them survive in the ~~desert.~~ *polar regions*

Penguins have adaptations to help them fly.

Desert animals need adaptations to help them to keep warm during the day.

The fennec fox has small ears to keep it cool.

Investigating habitats

Discover that different animals live in different habitats and how animals are adapted to their habitat.

The Big Idea

There are many different habitats and many different adaptations

Look at this picture of a leopard attacking a porcupine.

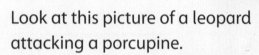 Discuss your observations with a partner.

Let's look at how some other creatures have adapted to survive in their habitat.

Look at this picture of a snake. The snake is going to eat the egg.

The jaws of the snake are adapted to move so that the snake can eat things that are bigger than its own head!

Can you identify any other adaptations the snake has to help it live in its habitat?

Look at these two snakes. One snake has venom and the other does not. But how would you know the difference? We need to be very careful when we see a snake because a snake bite can be very dangerous.

Eastern coral snake

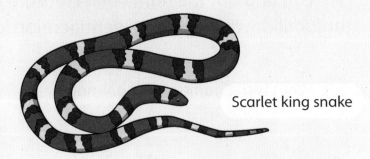

Scarlet king snake

Look at these pictures of animals that have adapted to live in their **environment**. Read the adaptations. Write the letter of each animal in the correct box.

a

Chameleon

b

Tortoise

c

Peregrine falcon

d

Puffer fish

1 I can make myself look very big to scare away other creatures. I produce a poison that can kill humans. ☐

2 I can change colour so I can hide from other creatures. I have a very long, sticky tongue. My eyes can move around to look for food. ☐

3 I move very slowly. I have a hard shell that protects me. I have claws to dig holes for my eggs. ☐

4 I have sharp talons to catch my food. I can dive through the air very fast. I have pointed wings. ☐

Write about other creatures you know. How are they adapted to live in their habitat?

Habitats

77

Now turn to page 94 to review and reflect on what you have learned.

Identification keys

Learn how to use identification keys.

The Big Idea

We can find out the name of a creature we are unfamiliar with by using identification keys.

✏️ What is the name of this animal?

We know what it is because we already know the name of this animal.

How can we find out the names of creatures we do not already know? We can use identification keys to help us! Look at this example of an **identification key**.

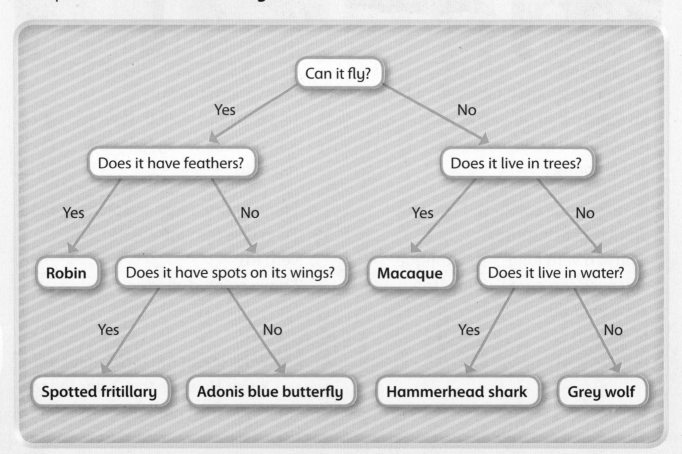

✎ Use the identification key on page 78 to identify these creatures.
Write the names of the creatures in the boxes.

1 _____

2 _____

3 _____

4 _____

5 _____

6 _____

✎ Make an identification key for these creatures.

Look at the pictures and think about how you can identify different features. Write the questions.

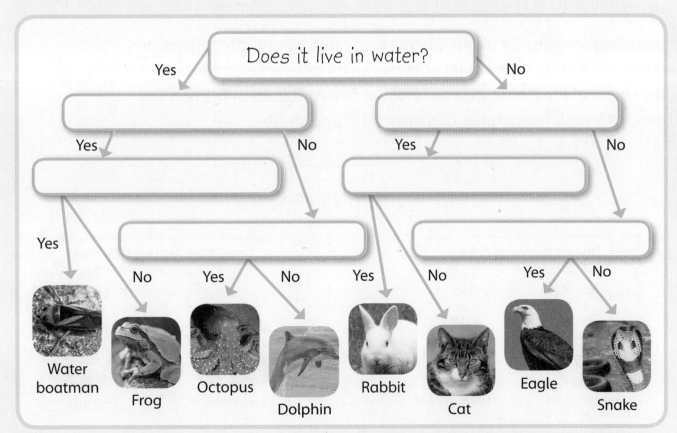

Identification keys

Learn how to use identification keys.

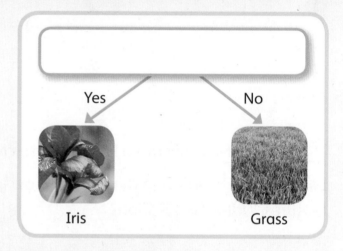

The Big Idea

We can find out the name of a plant we are unfamiliar with by using identification keys!

💬 Look at this plant. What is it called?

We can also use identification keys to help us to name plants that we do not know the names of. Let's try a very simple plant identification key to start with.

✏️ Can you write a question for this identification key?

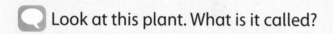

Yes   No

Iris   Grass

Identification keys for plants can become quite complicated. There are many plants that look similar but have small differences, for example the shape of the leaf or the colour of the flower. This identification key looks at two different types of plants: trees and shrubs.

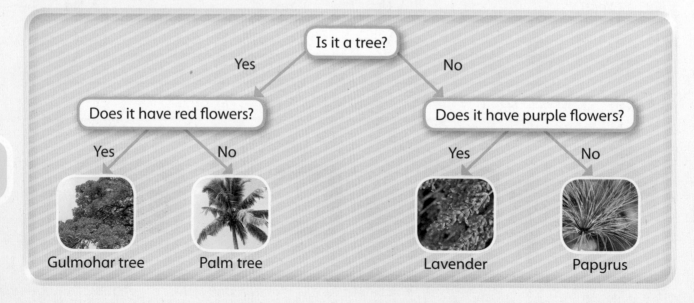

Is it a tree?

Yes   No

Does it have red flowers?   Does it have purple flowers?

Yes   No   Yes   No

Gulmohar tree   Palm tree   Lavender   Papyrus

The next pictures are of different varieties of one type of plant: cacti.

Identification keys for one type of plant are very useful to people who grow and sell flowers and plants. They need to help their customers find the plants that they want.

Imagine that you want to buy two very special cacti for a friend. You know that your friend likes flowering cacti with white flowers and round cacti with red spines.

Complete the identification key to help you find the right cacti for your friend.

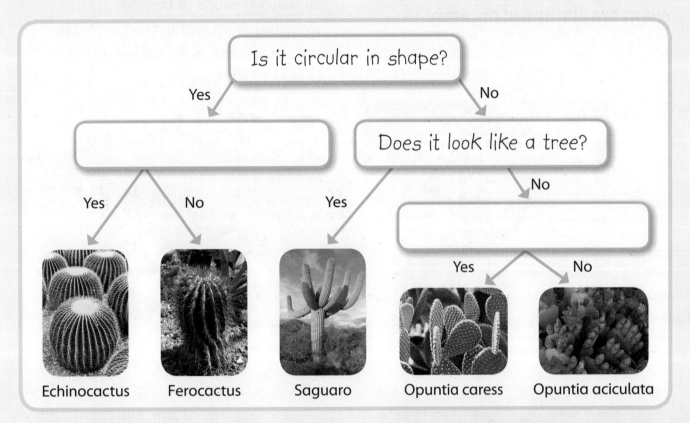

Is it circular in shape?

Yes

No

Does it look like a tree?

Yes   No

Yes

No

Yes   No

Echinocactus   Ferocactus   Saguaro   Opuntia caress   Opuntia aciculata

True or false?

1 We can use identification keys to help us name unfamiliar creatures.   (True) False

2 Identification keys can be simple or complex.   True False

3 We can only use identification keys to identify plants.   True False

4 Plants of the same type may look very different.   True False

Now turn to page 95 to review and reflect on what you have learned.

Habitats

81

How we affect our world

Find out how human activity affects the environment.

The Big Idea

We need oil, but if it is spilled it can harm natural habitats.

✎ Can you list two things that we use oil for?

Oil takes millions of years to form naturally deep within the Earth. The map shows the main oil-producing regions of the world.

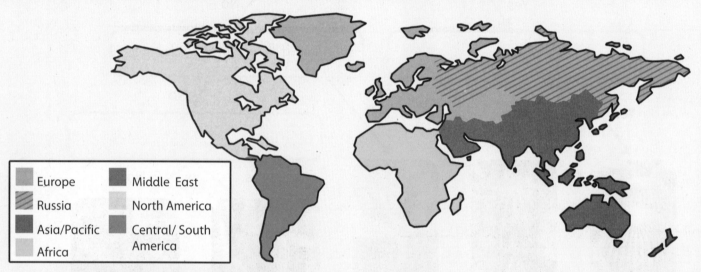

Europe
Russia
Asia/Pacific
Africa
Middle East
North America
Central/ South America

✎ Copy the bar chart in your notebook and use the information in the table to complete it. Then create another bar chart for **Oil production**.

	Oil reserves (billion barrels)	Oil production (million barrels per day)
Middle East	742.7	25.1
Europe/Russia	140.5	17.5
Africa	114.3	9.8
Central/South America	103.5	7
North America	59.5	13.6
Asia/Pacific	40.2	8

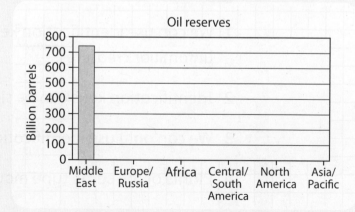

Oil is found both under the sea and under the ground. People have to drill into the earth to reach the oil. Companies send the oil to refineries and factories where it is made into useful products. They have to transport the oil carefully so that it does not spill.

Sometimes things can go badly wrong. Large oil spills cause big problems.

If oil is spilled in the sea, it makes an oil slick that can travel many kilometres across the oceans. It harms birds, fish and other animals that live in the sea. Creatures are covered in oil. They have

to be captured, cleaned and returned to their natural habitat if possible.

When the oil slick reaches land, it pollutes the beaches and damages natural habitats.

Think about...

How can we prevent oil spills? Think about how oil is drilled and how oil is transported.

 How are birds cleaned when they are covered in oil? Look at the pictures and write the numbers 1–6 in the boxes to show the correct order.

Drying

Preparation

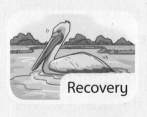
Recovery

Cleaning

Release

Rinsing

How we affect our world

Find out how human activity affects the environment.

The Big Idea

Tsunamis are natural disasters. We cannot stop them, but we can reduce how they affect the environment and places where people live.

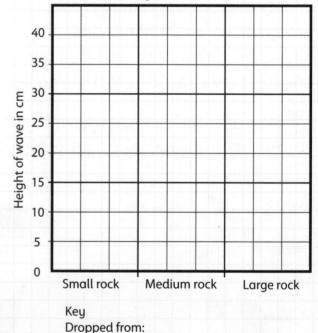

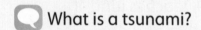 **What is a tsunami?**

A tsunami is a natural event. We cannot prevent tsunamis but we can prepare for them.

A tsunami is formed when there are movements of the Earth's surface under the oceans. These movements include underwater **volcanoes**, landslides and **earthquakes**. Very occasionally a tsunami is caused by a meteorite striking the Earth.

These movements create waves that move in all directions.

 Investigation: Investigating the size of waves

Watch the demonstration of the size of waves formed by dropping rocks into the water.

 Use the results you collected during the demonstration to create a bar chart.

Size of waves formed when dropping rocks from different heights

Height of wave in cm

40
35
30
25
20
15
10
5
0

Small rock   Medium rock   Large rock

Key
Dropped from:
☐ 50 cm   ☐ 100 cm   ☐ 150 cm

Q What does the bar chart tell us? Discuss the results with the class.

✏️ Then complete the sentence.

The bigger the rock, the _____

_____.

Preparing for tsunamis?

We have seen how tsunamis are formed, but what can we do about them? Many people can be hurt or killed when large tsumanis hit coastal areas. Large tsunamis can also cause a lot of damage to the environment.

✏️ How can we prepare for tsunamis? Use these suggestions to create an action plan.

- Move away from the coast quickly.

- Have an early warning system in places where tsunamis are likely to occur.

- Arrange where you will meet your family and friends if you are separated from them.

- Make your way quickly to higher ground.

- Make sure you know how to contact your family if there is a tsunami warning.

- Make sure you follow any instructions given.

Q How can we protect habitats from the effects of tsunamis?

✏️ True or false?

1 Tsunamis are always caused by meteorites.   True (False)

2 There is a link between the size of the tsunami and the force that created it.   True   False

3 A tsunami is a natural event.   True   False

4 We can stop tsunamis from happening.   True   False

How we affect our world

Find out how human activity affects the environment.

The Big Idea

We need clean air!

Look at the picture.

Do you think this air is clean?

Why is it important for us to have clean air to breathe?

We all need air to breathe and to live. Animals also need air to breathe and to live. If there are high levels of air **pollution**, the air we breathe can cause harm to us and to other creatures that breathe it.

Where does air pollution come from? Look at the picture. Write in the boxes using words from the word bank.

1
2
3
4
5
6
7

Word Bank

cars   animals   factory   power station   landfill site   trains   homes

We can measure air pollution. Some environmental scientists work in air pollution monitoring stations like this one.

Monitoring stations collect data all day every day. Scientists give air pollution a number from 1 to 10.

Look at the air quality index bands to see how air quality is shown.

1 2 3	4 5 6
Low	Moderate

7 8 9	10
High	Very high

If air pollution levels are very high, warnings are given.

We can do lots of things to reduce air pollution. Here are some suggestions.

Polluter	Suggestion
Factories	Use clean energy from renewable sources.
Cars	Walk or use hybrid cars.
Landfill sites	Reduce waste.
Power stations	Use clean energy from renewable sources.

Use these suggestions to create a poster about reducing air pollution.

Think about...
Is there any air pollution where you live?

Look at these air quality index values and decide if air pollution is low, moderate, high or very high.

Air quality index	Air pollution
5	Moderate
10	
1	
9	
3	

How we affect our world

Find out how human activity affects the environment.

The Big Idea

The ash and gases released from volcanic eruptions can reach heights of over 30 kilometres above the surface of the Earth!

This photograph shows the volcano Mauna Loa in Hawaii. It is the biggest volcano in the world.

What is happening in the photograph?

Volcanoes are one of the most dramatic natural things in our world. There are three classes of volcanoes.

Active   Dormant   Extinct

What is a volcano?

A volcano is an opening in the Earth's surface. Let's look inside a volcano.

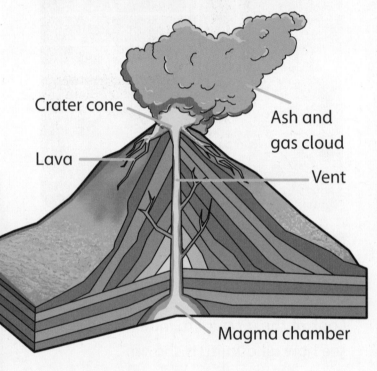

Crater cone

Lava

Ash and gas cloud

Vent

Magma chamber

The magma chamber is where an eruption starts.

Magma is very hot liquid rock. When pressure in the magma chamber builds up too high, it needs to escape. This is when eruptions happen.

The magma is forced up the vent very quickly and erupts from the crater. When the magma leaves the volcano it is called lava. The erupting volcano also releases ash and gas clouds.

Investigation: Making a volcano!

You will need some modelling clay, a small container and a large tray.

1 Place your container in the middle of the tray. Use the modelling clay to create the shape of a volcano. Leave the top of the volcano open for the eruption!

2 You need two tablespoons of baking soda and some vinegar with red food colouring in it.

3 Put the baking soda inside the container in your volcano. Add a little warm water and a few drops of washing up liquid.

4 Then pour the red vinegar into the container. Watch what happens.

How do human activities affect volcanic regions?

Many tourists find volcanoes fascinating because of the unusual landscape and the features.

Can you think of any problems that large-scale tourism creates in volcanic regions?

Use the words in the word bank to complete the paragraph about volcanoes.

Volcanoes form very dramatic _landscapes_. This attracts tourists. Volcanoes erupt when pressure in the _____ chamber gets too high and forces the magma through the vent. Magma is very hot liquid _____. Volcanoes are classified as active, _____ or extinct.

Word Bank

dormant ~~landscapes~~ magma rock

How we affect our world

Find out how human activity affects the environment.

The Big Idea

Human activities affect rivers in many ways.

Look at the picture of people enjoying spending time on a river. Do you know what they are doing?

Rivers come in many different shapes and sizes. Some rivers are wide and slow moving, others are narrow and fast flowing. Some rivers are very long, some rivers are very short.

What are the names of the three longest rivers in the world?

Humans use rivers for many different reasons.

How many activities can you think of that people do on or near a river?

We know that rivers are very useful for different reasons, but do we damage rivers by using them? Let's look at how human activities can impact on a river and its surrounding habitats.

Here we can see factories at the side of a river.

How do you think factories can have an impact on the river?

These pictures show different human activities on or near to a river. For each of the pictures identify the impact on the river and the surrounding habitats.

 Write the letter of the correct impact next to the photograph of each activity.

1

2

3

4

5

a Change the way the river flows

b Noise pollution

c Fewer fish

d Damage plants on the river banks

e Creatures eat the rubbish and become ill

 True or false?

1 Humans use rivers for many different activities.   (True) False

2 Humans use dams to create energy.   True False

3 Chemicals from factories can harm habitats near rivers.   True False

4 The Amazon is the world's longest river.   True False

Think about...
How do you think we can protect rivers and the habitats near to them?

How we affect our world

Find out how human activity affects the environment.

The Big Idea

The Earth can move and we cannot stop it.

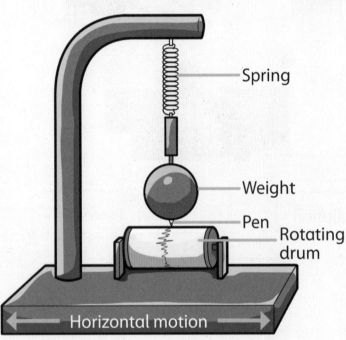

Spring
Weight
Pen
Rotating drum

Horizontal motion

Look at this picture of a model seismograph. What do you think it measures?

Earthquakes are **natural disasters** and we cannot stop them from happening.

Imagine that the Earth's surface is like a huge jigsaw with pieces that do not quite fit together. The scientific name for these jigsaw pieces is tectonic plates.

Sometimes the tectonic plates move and cause earthquakes.

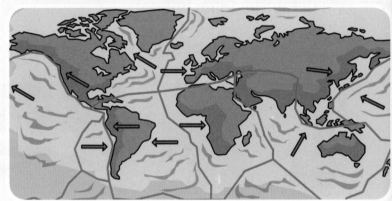

Scientists use seismographs to measure the movement of the Earth. They can give early warnings when there is going to be an earthquake.

The strength, or magnitude, of an earthquake is measured using the Richter scale of 0–10.

The biggest earthquake ever recorded was on May 22 1960 near Valdivia in Chile. It had a magnitude of 9.5.

 Create a bar chart in your Investigation Notebook showing the magnitude of these five earthquakes.

Valdivia, Chile (1960)	9.5
Prince William Sound, Alaska (1964)	9.2
Sumatra, Indonesia (2004)	9.1
Ecuador and Colombia (1906)	8.8
Northridge, USA (1994)	6.7

Earthquakes can destroy buildings, and people can be trapped under the fallen buildings. Fires often start.

We cannot prevent earthquakes from happening, but we can be prepared for them. We can design buildings that are earthquake-proof.

What do you notice about these buildings?

Burj Khalifa, Dubai

Transamerican Pyramid, San Francisco

Think about...
Can you design your own earthquake-proof building?

Draw lines to match the pictures with the correct words.

1

air pollution

2

earthquake

3

water pollution

4

oil spill

5

tsunami

6

volcano

Habitats

93

Now turn to page 95 to review and reflect on what you have learned.

Investigating habitats

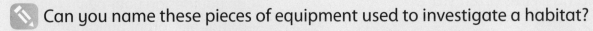 Can you name these pieces of equipment used to investigate a habitat?

1 | 2 | 3 | 4

Name an animal that is adapted to live in a cold habitat.

Name an animal that is adapted to live in a hot dry habitat.

Name one way this animal is adapted to the cold habitat.

Name one way this animal is adapted to the hot dry habitat.

 I know how to investigate the plants and animals in a habitat.

 I know some ways that different animals are adapted to their habitats.

Identification keys

 Here is a simple identification key. Which two words are missing? Write them in the correct places on the identification key.

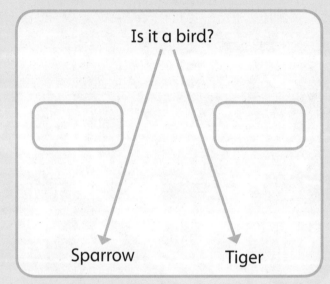

Is it a bird?

Sparrow   Tiger

I know how to use an identification key to identify animals and plants. ◯

I can design my own identification key. ◯

How we affect our world

 Name two types of pollution.

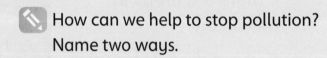

 How can we help to stop pollution? Name two ways.

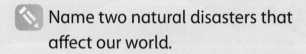

 Name two natural disasters that affect our world.

I know how humans can damage our world.

I know some ways we can help reduce the damage to our world.

Habitats

95

5 Making Circuits

In this module you will:

- construct your own electric circuits

- explore what happens if there is a break in the circuit

- find out that electrical current flows.

Word Cloud

make

component

circuit

battery

construct

flow

wire

design

electricity

Think about...

How does electricity work? Where does the electricity come from? How does a bulb give us light?

Amazing fact

We don't know exactly when electricity was first discovered. But we do know that the ancient Greeks knew about electricity over 2000 years ago.

Electricity – what a good idea!

Q What do we use electric lights for?

Q Imagine we live in a world with no electricity.

With a partner, think of three things that are different in this world.

Q List your favourite games and pastimes. How many of them use batteries? Would you swap them for this toy?

97

Constructing circuits

Construct your own electric circuits.

The Big Idea

We can use components to make our own electric circuits.

Draw lines to match the **components** with the words.

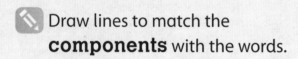 1

battery

 2

lamp

 3

wire

 4

switch

Electricity flows through the components in a **circuit**.

Can you list any other electrical components? Think about things you use at home.

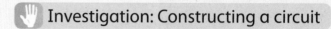 Investigation: Constructing a circuit

1 Can you **make** a circuit using the components in the pictures on this page?

2 Predict what will happen to the lamp when the circuit is complete.

This is a simple series circuit. We can use this kind of circuit to test components.

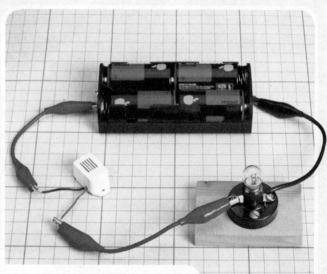

How can we find out why this circuit does not work?

We can use the test circuit to find out which component is broken. Take each component out of the circuit and put it in the test circuit one at a time. If the lamp in the test circuit lights, the component is working. If it does not, then you have found the faulty component that needs replacing.

Your teacher will give you some components to test.

1 First construct your test circuit and make sure that it is working.

2 Change the **battery**.

   If the lamp lights, the battery is working. If the lamp does not light, the battery is faulty and needs replacing.

3 Now test the other components. Test the **wires** and the lamps.

To make electricity flow we need a source of electricity. In your investigations, you will use a battery.

Batteries make electrical energy when chemicals inside them react. There are different parts in a battery called cells. They release tiny particles which move from the + side to the – side of the battery. When a wire is connected to the battery the particles move through the wire and back to the battery. As soon as the battery is connected the electricity flows until all the chemicals are used up.

Why are rechargeable batteries useful?

Make a list of three things that use a rechargeable battery.

Think about...

What do you think the + and – signs mean? Why are these signs on the battery?

⚠ Did you know that rechargeable batteries can be very dangerous?

They release the electricity very quickly and this can burn us.

Constructing circuits

Construct your own electric circuits.

The Big Idea

We can add more components to a circuit.

In this unit, you will **construct** circuits with more components.

Investigation: What happens when we add more lamps?

Do you think you can add more lamps to a circuit? Where will you put the lamps?

What do you predict will happen when you add more lamps?

I predict that _____

_____

Work with a partner. Investigate how many lamps you can add to your simple series circuit. Each time you add another lamp, observe how bright the lamps are.

Copy the table in your Investigation Notebook. Use it to record your results.

Number of lamps in circuit	Observation
1	The lamp is very bright.

Discuss with your partner what happened when you added more lamps to the circuit.

Complete the sentence about your results. Tick the correct box.

When we added more lamps, the lamps …

got brighter ☐

got less bright ☐

stayed the same ☐

What can we do to make the lamps brighter?

✋ **Investigation: What happens when we use more batteries?**

1 Repeat the investigation, but this time, use three batteries instead of one.

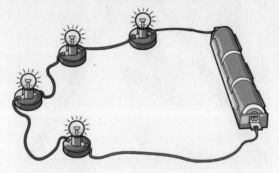

2 Make a table in your Investigation Notebook to record your results.

💬 Discuss with your partner what you observed. Compare your results with the first investigation.

✏️ Complete the concluding sentence.

We made a circuit with three batteries. When we added more lamps the lamps _____.

✋ **Investigation: What happens when we change the positions of the components?**

1 Construct the circuit in the picture.

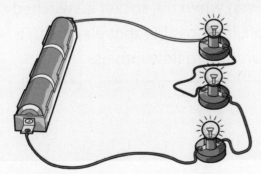

Observe how bright the lamps are.

2 Now construct the circuit in this picture. Does the brightness of the lamps change?

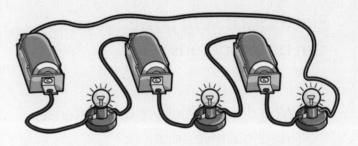

✏️ Circle the correct words in this sentence.

The position of the lamps and batteries changes / does not change the brightness of the lamps.

Think about...

Fairy lights are joined in a very long circuit. If one lamp stops working none of them work.

Why does this happen?

Constructing circuits

Construct your own electric circuits.

The Big Idea

Electricity is very powerful and can be dangerous.

Q Why do you think we use batteries in school investigations?

Batteries have a much lower voltage than mains electricity. The force of the electricity in a circuit is measured in volts. In most countries mains electricity has a voltage of between 220 and 240 volts.

Look at the batteries you have been using. Can you see what the voltage is? It is much lower than 240 volts. This means that batteries are much safer to use.

- Small things such as torches and toys don't need much electricity so they use batteries.

- Large things such as fridges and televisions need a lot of electricity so they use mains electricity.

⚠ Mains electricity is very dangerous. It can flow through us easily. This could kill us.

Have you seen this symbol before?

⚠ This symbol means high voltage. If you see this symbol you must stay away from it.

⚠ Do not touch an electric socket. Even when the socket is switched off, it is possible that electricity is waiting to flow into us.

> ⚠ Electricity can flow through metal things. Never stick scissors or spoons or any metal objects into a mains socket.

Electricity can also flow through water.

💬 Why should we never touch a socket with wet hands?

Look around the classroom. Where are the electrical sockets? Are there any sinks or taps close to them?

✏ Make a sign to warn people of the dangers of electric sockets.

The plastic part of a plug stops electricity flowing into us. Always hold this part when you are plugging or unplugging appliances. Wires are also covered in plastic to stop electricity flowing through us. If you see a break in the covering you should not use the appliance.

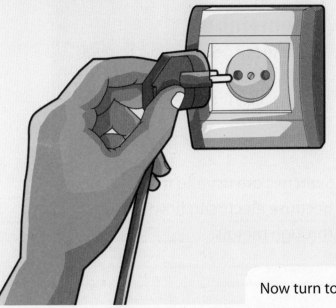

💬 What could happen if we use the appliance in the picture?

✏ Design a safety poster to display in your school.

Include all the rules from this unit. Explain why electricity is dangerous and why people need to be very careful.

✏ Answer the questions about electricity.

1 List five things that use mains electricity.

2 List three things that use batteries.

3 Mains electricity is dangerous because ...

Now turn to page 110 to review and reflect on what you have learned.

Making Circuits

Break in the circuit

Explore what happens if there is a break in the circuit.

The Big Idea

We can use a switch to control the flow of electricity.

💬 Why do we use switches to turn electricity on or off? Why don't we leave everything switched on?

We use switches in circuits to control the flow of electricity. It would be very dangerous to leave all the lights and appliances switched on all the time. It would also be very expensive. Switches allow us to use appliances when we want them on.

How does a switch work?

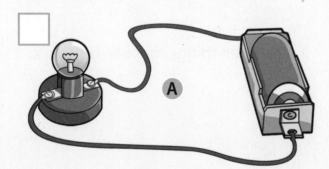

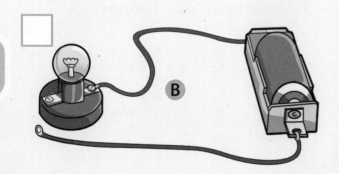

✏️ In which circuit will the lamp be lit? Put a tick next to the correct picture.

💬 Can you explain your answer? Why does the lamp not light in the other circuit?

Remember
If there is a break anywhere in a circuit, the electrical current will not flow.

💬 Can you think of any other ways the circuit could be broken?

We use a switch to control a circuit.

● When the switch is open, the circuit is broken and electricity cannot flow.

● When the switch is closed, the circuit is complete and electricity can flow.

Remember
Electricity needs to flow through something. If there is a break in the circuit the electricity will stop at the break.

Switches are usually made of metals because electricity flows easily through metals.

Investigation: Making a switch

You can make a switch.

You will need a small piece of cardboard, two brass paper fasteners and a paperclip.

1. Construct a simple circuit with a battery, wires and a lamp.

2. Connect the wires to the brass fasteners.

   This is an open switch. It makes a break in the circuit so the electricity cannot flow.

💬 What do you notice about the lamp in your circuit when the switch is open?

3. Slide the end of the paperclip to connect to the other brass fastener.

   This is a closed switch. The switch closes the break and completes the circuit. This allows the electricity to flow.

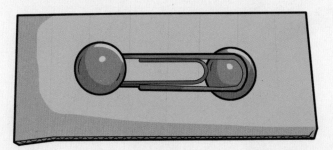

💬 What do you notice about the lamp when the switch is closed?

We can make a switch from any material that allows electricity to flow.

4. Try using foil, a nail or a pin to make a switch.

5. Connect the circuit. Take out the paperclip and replace it with other metals.

✏️ Which material worked the best?

The light switches you use at home work in the same way as the switch you have made. Can you describe how a light switch works?

Break in the circuit

Explore what happens if there is a break in the circuit.

The Big Idea

You can construct a circuit with a buzzer to make a doorbell.

💬 Will this circuit work? Why?

How can you fix it?

⚠️ Never use a wire that is broken or that has cracked plastic. Otherwise the electricity might flow into you.

✋ Investigation: Constructing a circuit with a buzzer

💬 How can you add a buzzer to your circuit? Where will you put the buzzer in the circuit?

1 In your Investigation Notebook, **design** a circuit with a buzzer. Draw a picture of the circuit.

2 Construct your circuit and test if it works.

If the circuit is complete, you can hear the buzzer. All the time! If you cannot hear the buzzer, the circuit must be broken.

💬 How can you find out what is broken in the circuit?

3 Now add a switch to your circuit, so you can switch the buzzer on and off.

You can use the switch that you made earlier. This is how a doorbell works.

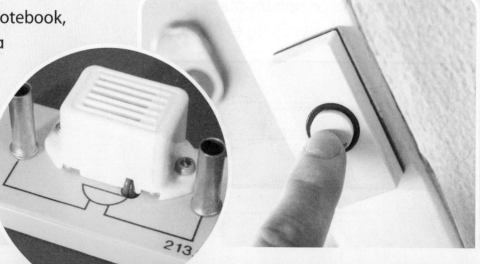

Investigation: Constructing different circuits

Look at the pictures of circuits.

1 In your Investigation Notebook write your prediction for each circuit.

2 Construct the circuits and test your predictions. Record your observations.

Think about... ?

Bells are often used instead of buzzers. Can you think of any uses for electric bells?

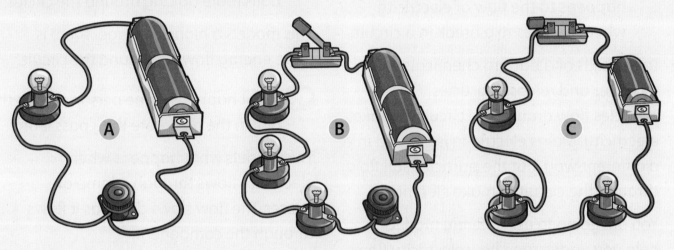

Complete the sentences about circuits by drawing a circle around the correct words.

1 When there is a break in the circuit the components
will work / will not work.

2 When the switch is open, the electricity can / cannot flow.

3 When the switch is closed the electricity can / cannot flow. This means the buzzer will work / will not work.

How can you make a lamp in a circuit brighter? Tick the correct answer.

Add more bulbs ☐   Add a switch ☐   Add more batteries ☐

Now turn to page 111 to review and reflect on what you have learned.

Making Circuits

107

Electrical current flows

Find out that electrical current flows.

The Big Idea

Electricity flows around a circuit.

💬 Explain to your partner what happens to the flow of electricity when it comes to a break in a circuit.

In the cells of a battery chemicals react together and release particles. The particles flow around the circuit to make electricity. Mains electricity is created in a different way, but the particles still flow through the circuit to make electricity.

You are going to use different models to help you understand how electricity flows.

Passing balls around a circle

Pass balls around a circle to represent electrical current flowing around a circuit. The balls are like the tiny particles that create electricity.

💬 What happens to the flow if someone drops the ball?

This models a break in the circuit. The current cannot flow.

💬 What happens if everyone passes the balls more quickly around the circle?

This models a higher voltage. There is more energy flowing around the circuit.

💬 What happens if one person throws the ball in the air before they pass it on?

This models what happens when electricity flows through a lamp or buzzer. The flow slows down as it flows through the component.

💬 What happens if you make a gap between two people in the circuit?

This models an open switch in a circuit. The flow stops when it reaches the break in the circuit.

Central heating model

We can model the flow of electricity by comparing it to a central heating system.

The parts of the central heating system are labelled on the picture.

 Which parts of an electric circuit do the components represent? Write the labels in the correct boxes.

Word Bank

battery   **electricity**   **lamp**   **wires**

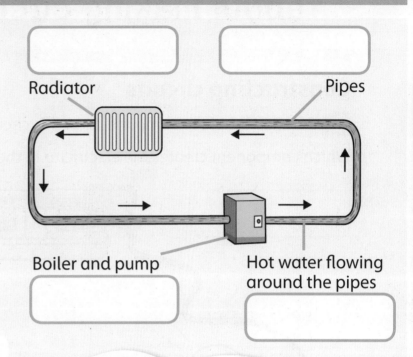

Radiator

Pipes

Boiler and pump

Hot water flowing around the pipes

Think about...

There are some problems with the central heating model. Can you see what the problems are?

Amazing fact

The tiny particles move around a circuit at the speed of a snail. The energy they carry moves at the speed of light.

 What have you learned from modelling electricity? Draw a line to match each description to the correct word.

1 This makes a break in the circuit so electricity cannot flow.

2 These flow around the circuit and create electricity.

3 This completes the circuit and electricity can flow.

4 This slows down the flow of electricity but does not stop it.

Particles

Component

Closed switch

Open switch

Now turn to page 111 to review and reflect on what you have learned.

What we have learned about making circuits

Constructing circuits

 Can you label the components in this circuit?

Which component creates the electricity in the circuit? Tick this component.

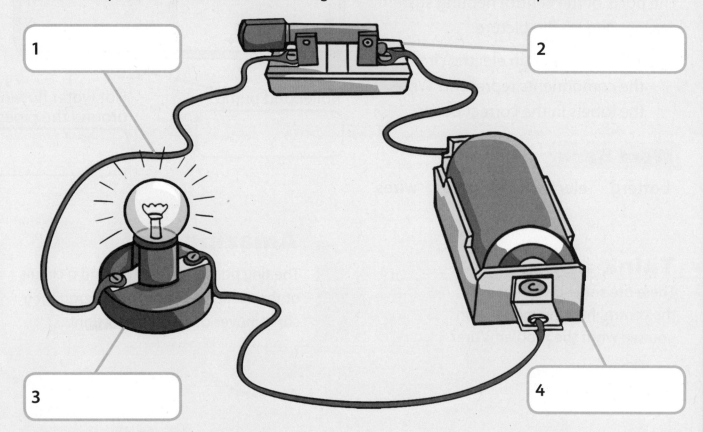

1

2

3

4

 How can you find a fault in a series circuit?

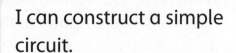 I can construct a simple circuit.

 Why is mains electricity much more dangerous than electricity from a battery?

I can test a simple circuit for a fault.

I know the rules for using electricity safely.

110

Break in the circuit

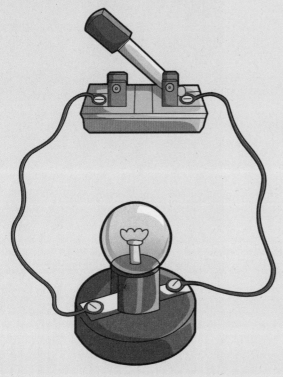

 Why doesn't the lamp light in this circuit?

Complete the sentence.

Electricity will only flow through a
_____ circuit.

I understand how a switch works. ◯

I know electricity flows around the circuit and back to the battery. ◯

Electrical current flows

Read the sentences about a model for electrical current. Write the correct words in the gaps.

We can pass _____ around a circle to model electrical _____ flowing around a _____. The balls are like the tiny _____ that create _____.

Word Bank

circuit   particles   balls

current   electricity

I know that an electrical current carries energy. ◯

I can use models to explain how the particles flow around a circuit. ◯

6 Sound

In this module you will:

- explore how sounds are made and learn how to measure sound

- investigate how sound travels through different materials to the ear

- investigate how some materials prevent sound from travelling through them

- investigate how high or low a sound is and that high and low sounds can be loud or quiet

- explore how we can change pitch to make musical instruments.

Word Cloud

quiet

volume prevent loud

vibrate material

decibel

travel pitch

Amazing fact

Dolphins hear through their jaws because they have tiny ears.

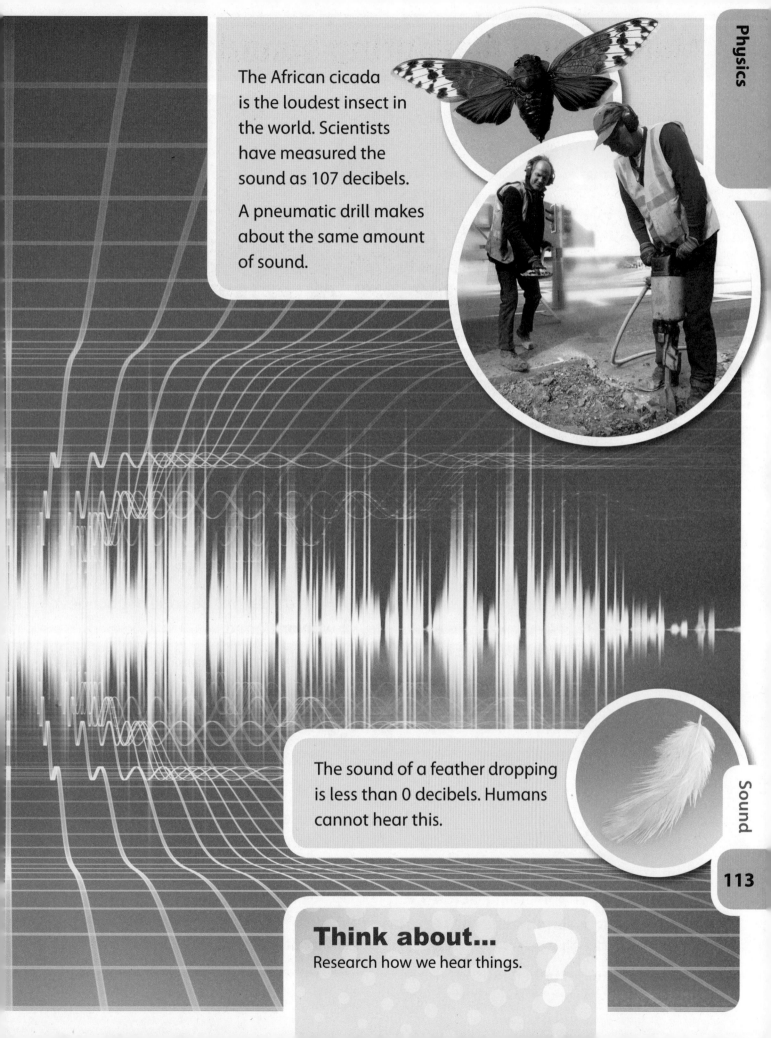

The African cicada is the loudest insect in the world. Scientists have measured the sound as 107 decibels.

A pneumatic drill makes about the same amount of sound.

The sound of a feather dropping is less than 0 decibels. Humans cannot hear this.

Think about...
Research how we hear things.

Making and measuring sound

Explore how sounds are made and learn how to measure sound.

The Big Idea

We hear sound when something vibrates.

💬 **What is sound?**

There are lots of sounds around us. We can make sounds by talking, clapping our hands and whistling. Cars, radios and TVs make sounds that we hear.

✏️ What can you hear? Make a list of all the sounds that you can hear now.

Sound can only happen if something **vibrates**. When something vibrates, it makes the air or surrounding **materials** vibrate too. The vibration moves across the air or material until it reaches the ear.

✋ **Investigation: We need vibrations to hear sounds**

1 You can demonstrate this.

   Hold a ruler on the edge of your desk with one hand.

2 With your other hand, pull the end of the ruler down and let it go.

💬 What happens to the end of the ruler?

   What happens when the ruler stops vibrating?

The loose end of the ruler vibrates. This vibrates the air around it. We hear a sound because the air vibrates in our ears. We can see the ruler vibrating up and down but we cannot see the air vibrating.

Investigation: How can we investigate the vibrations that cause sound?

1 Hit a drum with a stick or your hand.

2 Put rice on to the drum and then hit it.

💬 What happens to the rice?

What causes this?

Musicians and singers use tuning forks. If you tap the end of the tuning fork it starts to vibrate.

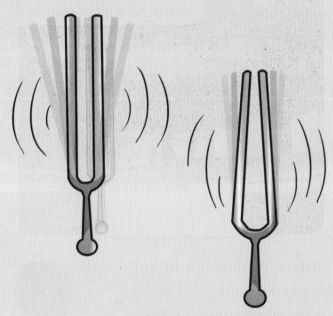

💬 How can we hear a sound from the tuning fork? How does the sound reach our ears?

Tap the tuning fork and put it on the rim of a container of water. What can you see on the surface of the water? This is the sound vibration moving through the water.

We hear a guitar because the string vibrates. The string vibrates the air as it moves to our ears. How can we investigate this?

Investigation: Make a guitar

Make a guitar like the one in the picture.

💬 What happens when you pluck the string? What does the vibrating string do to the air? How does this make us hear the sound?

Think about...

How can you make your guitar sound louder or quieter?

Making and measuring sound

Explore how sounds are made and learn how to measure sound.

The Big Idea

We measure the volume of sound in decibels (dB).

How do we know how loud a sound is?

How did you make your guitar sound louder?

The **volume** of a sound is how **loud** or **quiet** it is. The more energy you put into the vibration, the louder the sound will be.

Investigation: Volume

1 Gently hit or pluck a musical instrument. Is the sound loud or quiet?

2 Now hit or pluck the instrument as hard as you can. What is the difference in sound?

Look at the photographs. Is each sound loud or quiet?

A   An aeroplane taking off

B   Leaves falling from a tree

Scientists measure sound using a sound-level meter. It measures accurately the vibrations in the air. Scientists measure sound in a unit called **decibels** (dB).

What is the reading on this meter?

Investigation: Sound levels in the school

1 Use a sound-level meter to measure the sound levels in your school.

2 In your Investigation Notebook make a table to record your results.

Place	Sound level (dB)
Canteen	65 dB

Which place do you think will be the loudest? And the quietest?

 Answer the questions about your investigation in your notebook.

1 What was the highest reading you took?

2 Why was this place so loud?

3 List three places that had a reading below 60 decibels (dB).

Humans cannot hear sounds that are below 0 dB. If a sound is too loud it can be very painful. Sounds above 160 dB permanently damage our ears.

Scientists have made a list of different sound levels. This list helps us keep our hearing safe.

⚠ Never listen to very loud sounds. They can hurt our ears.

Sound	Sound level (dB)
Rustling leaves	10
Whisper	20
Conversation	60
Busy traffic	70
Vacuum cleaner	80
Music through headphones	100
A child screaming	110
Causes humans pain	130
Jet taking off	140
Permanent damage to the ear	160

✎ True or false?

1 Sound is made when something vibrates. (True) False

2 Tuning forks measure sound. True False

3 We measure sound in a unit called metres. True False

4 Very loud noises can damage our ears. True False

Now turn to page 132 to review and reflect on what you have learned.

How does sound travel to our ears?

Investigate how sound travels through different materials to the ear.

The Big Idea

Sounds need something to travel through for us to hear them.

💬 How does sound travel to our ears?

Sound can **travel** through materials and air. When an object vibrates it makes the air or any material next to it vibrate also.

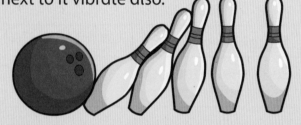

It is like playing skittles. When we hit one skittle with a ball, it knocks into the next skittle and that one knocks into the next skittle. Sound vibrations travel through materials and air in a similar way.

Sound can travel through lots of different materials.

💬 Can sound travel through a wall?

✋ Investigation: Does sound travel well through all materials?

1 Copy the table in your Investigation Notebook.

2 Make a list of all the objects you can test and the materials they are made of.

For example: stand with your ear against the wall. If you listen carefully you can hear sounds. This means that the vibrations are travelling through the wall and vibrating in your ear. The wall is probably made out of brick or stone. This means sound can travel through the brick and stone.

Object	Material	Sounds I heard
Wall	Bricks	Talking, voices and children playing
Window	Glass	
Door	Wood	
Curtain	Cotton	

3 Test all the objects in your list. Record your results in your table.

118

✏️ Complete the sentences about your investigation.

It was easiest to hear through the _____. It is made of _____. This means that sound vibrations travel well through _____.

It was hardest to hear through the _____. It is made of _____. This means that sound vibrations do not travel well through _____.

Remember
Sound needs a material to travel through.

What happens to sound in a vacuum?

Bell jar

Telephone

Vacuum pump

In a vacuum there is no material or air for vibrations to travel through. The photograph shows a bell jar. A pump sucks all the air out of the jar. The phone is ringing but we cannot hear it. This is because there is nothing to transmit the vibrations from the phone to our ears.

Think about...
Where do the vibrations go?

?

Space is a vacuum, just like inside the bell jar. The astronaut in the photograph is working on a space ship.

💬 Can he hear the sound he makes as he repairs the space ship?

Will the people inside the space ship hear the sound?

How does sound travel to our ears?

Investigate how sound travels through different materials to the ear.

The Big Idea

Sounds change when they travel through different materials.

 Investigation: Which materials transmit sound the best?

1 Work with a partner.

2 One person puts their ear against an object and the other whispers a word against the object.

If you hear the correct word clearly, the material has transmitted the sound well.

3 Repeat each test to give you more reliable results.

Material	Correct word heard
Glass window	✓

 Which material transmitted sound the best?

 What does your investigation tell you about materials? Complete the sentences.

Some materials, for example _____, transmit sound well.

Some materials, for example _____, do not transmit sound well.

This proves that not all materials transmit sound in the same way. This is because sound vibrations move differently through different materials.

What happens to sounds under water?

Have you ever swum under water? Did you notice that sounds seem different? Sound travels up to five times faster under water than in air. We hear the sound differently because the sound moves differently in water than in air. Our ears become filled with water and so the vibrations in our ears are different.

120

Remember

Sound needs a material to travel through. Some materials are much better transmitters of sound than others.

We can use this knowledge to make quiet sounds seem different.

✋ Investigation: How does water change the way we hear sound?

1 Gently blow over the neck of an empty bottle. Can you make a sound?

The air from you is vibrating the air in the bottle. This is vibrating the air back into your ears.

2 Add some water to your bottle and gently blow over the neck like you did before.

💬 What happens to the sound now?

3 Add different amounts of water and investigate the sound you make.

💬 Is there a pattern?

✏️ What do you conclude from your investigation? Circle the correct words to complete the sentence.

When I added more water the volume of the sound got louder / got quieter / stayed the same and the sound got higher / lower.

Think about...

Why do some materials transmit sound better than others?

Hint: Think about what you know about particles in different materials.

How does sound travel to our ears?

Investigate how sound travels through different materials to the ear.

The Big Idea

Some materials help us to transmit sound so that it is louder and travels further.

💬 Can sound travel through all materials?

✋ Investigation: Which materials make sounds louder?

Can you make the tuning fork sound louder?

1 Hit the fork on a hard surface and hold it in the air. Listen to the sound it makes.

2 What happens to the sound when you put the handle on different surfaces? Is it louder or quieter?

3 Copy the table and record your observations.

Material	Observation
Desk	The sound was louder than in air.

4 Write your conclusions in your Investigation Notebook.

💬 How can you make your voice louder?

Try talking through a tube. Take it in turns to listen to your partner's voice with and without the tube.

⚠️ Do not shout into the tube. It could damage your partner's ears.

✏️ Does your partner's voice change when they talk through the tube?

The tube stops the vibrations escaping into the surrounding air. It keeps all the vibrations in the tube and transmits them straight into your ear. This makes the sound louder.

The vibrations travel in just one direction through the tube. This makes the sound travel further.

✋ Investigation: Make a telephone

Make a telephone from two paper cups and a long piece of string.

💬 Can you hear your partner speak?
Can you hear them whisper?

This is how the telephone works:

1 Your vocal cords cause waves.
These vibrate the air particles.

2 The vibrating air particles make your cup vibrate.

3 The vibrations from the cup vibrate the string.

4 The vibrations from the string vibrate your partner's cup.

5 The cup vibrates the air, then your partner's ear drum.

✏️ Name three things that sound can travel through.

✏️ Explain how we hear sounds. Use the words in the word bank to help you complete the sentences.

Vibrations from a sound _____.
_____ to our ears.

Word Bank

travel   air   material   transmits   sound   vibrations

✏️ Choose the correct word to complete this sentence.

The more energy there is in a vibration, the quieter / louder the sound.

Now turn to page 132 to review and reflect on what you have learned.

Sound

123

Some materials stop sound travelling

Investigate how some materials prevent sound from travelling through them.

The Big Idea

We can use some materials to protect our ears from very loud sounds.

An aeroplane taking off has a sound-meter reading of 140 dB. Sounds of 130 dB hurt our ears and sounds of 160 decibels dB cause permanent damage.

💬 How can people who work at an airport protect their ears from damage?

Some materials are better at transmitting sound than others. To **prevent** damage to our ears from loud noises we use materials that are not good at transmitting vibrations.

✋ Investigation: Which material makes the best ear defenders?

You will need a radio, a box and some materials to test.

1 Switch on the radio and place it in the box. Can you hear the radio?

 What piece of equipment could you use to measure the sound?

2 Wrap the box in each material and take a sound-meter reading.

If you do not have a sound-level meter, use a scale of 1 to 10 for how well you can hear the radio.

3 Copy the table in your Investigation Notebook and record your results.

Material	Sound level recorded (dB)
Cotton wool	60 dB

✏️ Which material will you use for your ear defenders?

✏️ Which material would make the worst ear defenders?

✏️ Can you explain why?

Sound insulation is not just used for ear defenders. Recording studios use insulation to prevent sound from outside travelling into the studio. Curtains, carpets and rugs are good insulators of sound. Some studios put carpet and rugs on the walls.

Think about...
Can you think of any other uses of sound insulation?

✏️ Answer the questions about sound insulation.

1 Explain why we need to lower the sound going into our ears.

2 List two materials that insulate sound.

3 Why do some airport workers wear ear defenders?

Sound

125

Now turn to page 132 to review and reflect on what you have learned.

Investigating pitch and volume

Investigate how high or low a sound is and that high and low sounds can be loud or quiet.

The Big Idea

 Sounds can be high or low.

Listen to a piece of music.

💬 How many different sounds can you hear?

What is the highest note you can sing? The **pitch** of a sound is how high or low it is.

✋ Investigation: How can we change the pitch of a sound?

Try the investigation with a ruler again.

1 Start with a long length of ruler vibrating.

💬 Is the pitch of the sound high or low?

2 Now hold the ruler so that only a small part of it vibrates.

💬 Is the pitch higher or lower?

3 Use your guitar to explore pitch. Change the length of a string by holding it down with one finger. Then pluck the string.

4 Make the string longer and shorter and listen to what happens to the pitch.

✏️ Use the words in the box to write two sentences about how to change the pitch of a guitar string. Use some of the words more than once.

Word Bank

guitar	string	long	short
pitch	low	high	

An oscilloscope shows both the pitch and the volume of a sound. It shows this as a wave pattern.

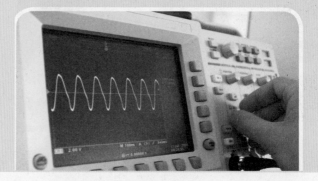

The girls are measuring the volume and pitch of their voices. The microphone picks up the vibrations from their voice. The oscilloscope turns the vibrations into a wave pattern on the screen.

1
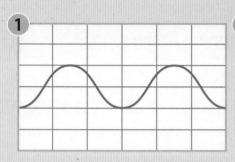

2
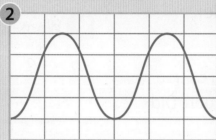

3
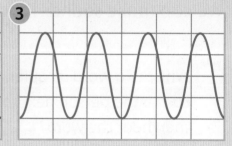

The number of waves tells us the pitch of a sound.

- In diagrams 1 and 2 there are only two waves. These waves are made by low sounds.

- In diagram 3 there are four waves. These waves are made by a high sound.

The height of the waves tells us how loud a sound is.

- In diagrams 2 and 3 the waves are high. These waves are made by loud sounds.

- In diagram 1 the waves are not as high. These waves are made by a quieter sound.

Look at the diagrams from an oscilloscope.

A
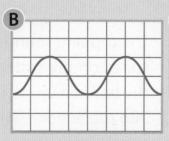

B
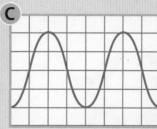

C

D
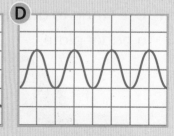

1 Which two are the loudest sounds?

☐ and ☐

2 Which sound has the highest pitch?

☐

Investigating pitch and volume

Investigate how high or low a sound is and that high and low sounds can be loud or quiet.

The Big Idea

 We can make sounds louder.

💬 Whisper a word as quietly as you can to the person next to you. Did they hear the word?

Remember

The more energy there is in the vibration the louder the sound. When we shout we put more energy into the vibration. When we whisper we put less energy into the vibration.

✋ Investigation: Make a tambourine and investigate how to play it quietly or loudly

1 Use string to attach bells or shells to a paper plate.

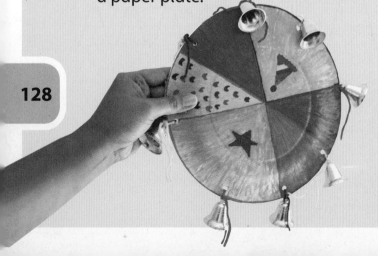

128

2 Hold the tambourine and gently tap the centre.

✏️ Describe what you hear.

3 Now hit it harder.

✏️ What happens to the sound?

Air is a good transmitter of sound. If you can force or push air particles together, sound travels even better. This makes the sound louder to our ears. How can we investigate this?

✋ Investigation: How can we amplify sound?

1 Blow up a balloon.

2 Place the balloon against your ear.

3 Gently tap on the balloon. Or ask someone to whisper into the balloon.

The tapping sounds very loud. This is called amplification. When you blow the balloon up you force air particles into the balloon. The particles of air are squashed together very closely. The vibration from your tapping can move through the balloon very easily and carry the vibration to your ear.

Q When you whispered to your partner, did you cup your hands around your mouth?

Q Why do you think we do this?

Make an ear cone like the one in the picture. Listen to some music. First with just your ears. Then with one cupped hand. Then with two cupped hands. Finally use your ear cone.

Q Which technique makes the sound louder? Explain why this happens.

✏ Answer the questions about pitch and volume.

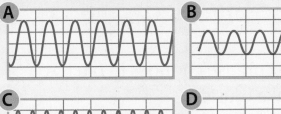

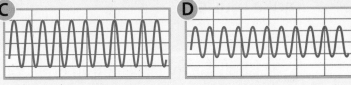

1 Which of these sounds have the highest pitch?

[ ]   and   [ ]

2 Which of the sounds are the loudest?

[ ]   and   [ ]

3 Write one way that you have amplified sound.

4 Explain how an amplifier works.

Now turn to page 133 to review and reflect on what you have learned.

Making music

Explore how we can change pitch to make musical instruments.

The Big Idea

We can use changes in pitch to make music.

List all the musical instruments you have made.

How did you change the pitch of the guitar you made?

Remember

If you change the length of the object that vibrates, the pitch changes.

When there was a small amount of ruler vibrating the pitch was much higher. We shortened the guitar string by pressing a finger on the string.

How do you think we can change the pitch of a drum?

In all instruments, if we can change the size of the object vibrating, we can change the pitch. When musicians play an instrument, they press on the string to change the length. This changes the pitch and makes a tune.

Investigation: Making a musical instrument with bottles

Bottles make good instruments. We investigated blowing on bottles. When we add water to the bottle the pitch changes. This is because there is less air to vibrate.

1 Fill bottles to different levels.

2 See if you can make a tune.

130

💬 What do you think will happen to the pitch if we tap the bottles with a pencil?

Does the pitch change in the same way as when we blow?

Explain to your partner what happens to the sound when you tap the bottle.

✋ Investigation: Making music with glasses

1 Run your finger around the rim of a glass.

The vibration from your finger goes into the air in the glass and makes a sound.

💬 What do you think will happen to the pitch if you add some water?

2 Set up six glasses and add different amounts of water to each.

3 Try to make a tune.

✏️ List all the instruments you have made now.

You could make an orchestra in your group.

✏️ True or false? Change the false sentences so they are true.

1 If the vibrating object is longer the pitch will be ~~higher.~~ lower   True (False)

2 If you blow over a bottle of water and then tap it with a pencil, it will have the same pitch.   True   False

3 The more water there is in the bottle, the lower the pitch.   True   False

Now turn to page 133 to review and reflect on what you have learned.

What we have learned about sound

Making and measuring sound

✏️ What is the unit of measurement of the loudness of sound?

[ ]

✏️ Give an example of where we can see the vibrations that invisible sound waves make.

[ ]

I know that sound is made by vibrations.

How does sound travel to our ears?

✏️ What happens to the particles in the air as sound travels?

[ ]

✏️ Why can we hear better in water than in air?

[ ]

I understand that sound travels to our ears by vibrating particles as it moves.

Some materials stop sound travelling

✏️ The particles in some materials stop sound travelling. Circle the materials that are good insulators.

bricks   bubble wrap   carpet

plastic   glass   foam   wood

curtains   metal   cardboard

How can you use these kinds of materials?

 What is the scientific word for making sound louder?

I understand the difference between pitch and volume. ○

Making music

1
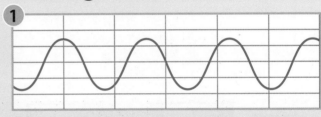

2
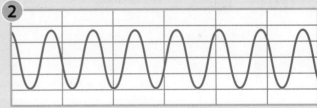

3
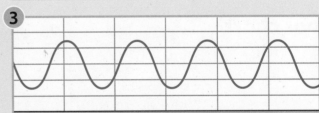

I know how to protect my ears and my home from loud sounds. ○

Investigating pitch and volume

 Tick the correct ending for this sentence.

The pitch of sound is …

how loud or quiet it is ☐

how high or low it is ☐

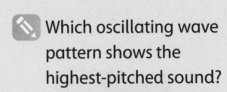 Which oscillating wave pattern shows the highest-pitched sound? ☐

I can change the pitch of sounds to make music. ○

Glossary

animal

battery
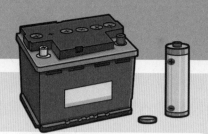

boiling

attract

bone

bar chart

circuit
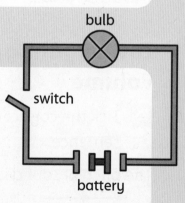

bulb

switch

battery

bar magnet
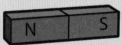

component

decibel

design

construct

electricity

danger

environment

equipment

freezing

gas

flow

habitat

force

identify

investigate

make

magnet

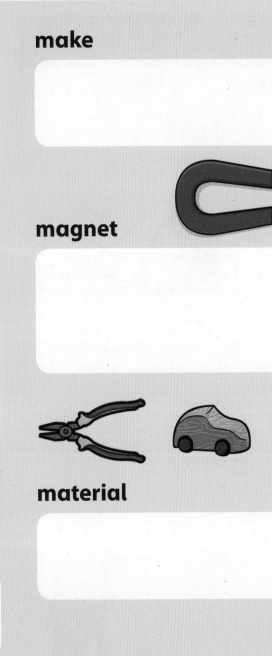

iron

material

liquid

matter

loud

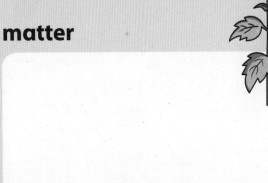

measure

muscle

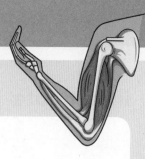

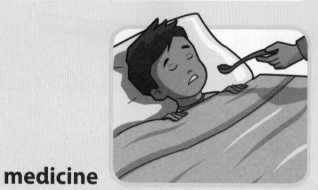

medicine

natural disaster

melt

North

move

observe

particle

pitch

pollution

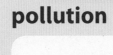

pooter

plant

predict

pole

prevent

quiet

record

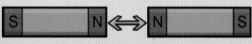

repel

skeleton

skull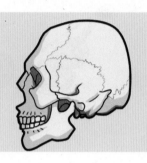

solid

South

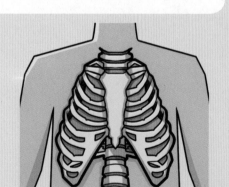

rib

spine

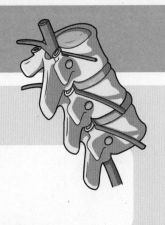

vibrate

steel

volcano

test

travel

volume

trend

wire